"Artificial intelligence is the defining force reshaping every profession and industry. The AI Transformation *provides an engaging, practical, and remarkably comprehensive guide to understanding this historic shift. Drawing on the expertise of 33 AI leaders, this book equips readers with the knowledge and perspective needed to thrive in an AI-powered world."*

—DAVID BROWNLEE, Bestselling Author of *Customer Service Success*

"Few books capture both the enormous promise and the real-world impact of artificial intelligence as effectively as The AI Transformation. *This collection delivers practical insights from experts who are applying AI every day. Whether you're a business leader, educator, or entrepreneur, this is an essential read."*

—ERICA WHARTON, Adjunct Professor of Change Management

"The greatest revolutions in history have transformed how humanity lives, works, and thinks. Artificial intelligence is proving to be no exception. The AI Transformation *offers readers a rare opportunity to learn directly from dozens of accomplished professionals who are shaping this new era. Thoughtful, accessible, and inspiring, this book belongs on every modern leader's bookshelf."*

—DAWNY PHILLIPS, AI Strategist and Consultant

AF872314

"Artificial intelligence is touching virtually every aspect of modern life, yet few resources explain its impact with such clarity and breadth. The AI Transformation *brings together an extraordinary collection of experts whose combined experience spans business, healthcare, education, engineering, cybersecurity, government, and beyond. The result is an insightful roadmap for anyone determined to succeed in the age of AI."*

—MOHAMMAD AL WAHEED, Founder of Ignite AI

"Technology alone has never transformed civilization, people have. The AI Transformation *reminds us that while artificial intelligence is changing the tools we use, human creativity and adaptability remain at the center of progress. This compelling collection of perspectives about AI will challenge your assumptions and prepare you for the opportunities ahead."*

—CASEY LINDBERG, Space X Quality Control Manager

"The AI Transformation *is more than a book about technology; it is a guide to one of the most profound shifts in human history. Featuring insights from 33 AI experts across a range of disciplines, it provides readers with the confidence and knowledge needed to navigate an increasingly intelligent world. Timely, balanced, and deeply informative, this book is a great introduction to AI."*

—MARIA GARCIA, Technology Advisor, Keynote Speaker

THE AI TRANSFORMATION

THE AI TRANSFORMATION

Thriving Within Civilization's Next Big Disruption

Authored by:
Erik Seversen, Amany Alkhayat, EdD, Scott Beal, Tanya J. Beck, Dr. De'Anjelo J.L. Bradley, William Brendel, PhD, Michal Buczko, Patricia Calazans, Jeffrey C. Granvold , Scott Hall, Ben Hamilton, PhD, Latif Hamlani, Michael Harris, Dr. Sheik M. Isaacs, Karlyn Kieffer, Vivek Kompella, Byron M. Lowens, PhD, Mark L. Madrid, Leonardo Veri, Nicolas Mendiharat, Yashaswini Nalla, Jeanette Okwu, Tomomichi Ono, Sridhar Peddisetty, Venkat (Ramana) Chivukula, Peter Rose, Ghina Sahagun, Mohammed Asif Sheroz Ali, Raymond Sondak, Jennifer L. Stapleton, Ivo Strohhammer, Sakina Syed, Gary Vause, II, and Rick Wilson

THIN LEAF PRESS | LOS ANGELES

Library of Congress Cataloging-in-Publication Data
Names: Seversen, Erik, Author, et al.
Title: *The AI Transformation: Thriving Within Civilization's Next Big Disruption.*
LCCN: 2026918614

ISBN 978-1-968318-74-1 (hardcover) | 978-1-968318-73-4 (paperback)
ISBN 978-1-968318-72-7 (eBook) | 978-1-968318-75-8 (audiobook)

Artificial Intelligence, Science & Technology, Business, Professional Development
Cover Design: 100 Covers
Interior Design: Dindo Sanguenza
Editor: Nancy Pile
Thin Leaf Press
Los Angeles

THIN
LEAF

Thank you for reading this book. There is information found within the following pages that can greatly benefit your life, but don't stop there. Make sure you get the most you can from this book and reach out directly to the expert-authors who want to help you navigate the AI transformation so you can thrive within civilization's next big disruption and manifest success in your life. Contact information for each author is found at the end of their respective chapter.

To those transforming the world with cutting edge technologies
aiming to create a positive future for all.

CONTENTS

INTRODUCTION

By Erik Seversen
Author of *Ordinary to Extraordinary*
Los Angeles, California

The future is not something we enter.
The future is something we create.
—Leonard I. Sweet

The word "transformation" can mean many things. The first that pops into my mind is a butterfly, whereby a quite disgusting larva transforms into something beautiful. Metaphors abound from this transformation from goo trapped inside a cocoon to a colorful creature capable of spreading its wings across continents.

Another transformation can be seen in the French fairy tale "Beauty and the Beast" whereby a cursed prince is first transformed into a hideous beast only to be transformed back into human form if he loves and gains the love of another. If you know the story, you may recall the wonderful feeling when this condition is met and the beast transforms back into human form, to the surprise and delight of Belle, the woman whose love allowed this transformation.

However, not all transformations are positive. In the 1962 film *Lawrence of Arabia*, we see the slow transformation of Lawrence as he changes from an idealistic, yet misfit, lieutenant in the British Army during the Arab Revolt of World War I, to a person capable of leading a massacre. In this story, this main character transforms from a naïvely innocent soldier into a barbarian who can't escape the horrors he witnesses.

While these are precise examples of transformation, transformations happen all around us all the time. From the yearly updates to automobiles, making them more efficient and functional to villages that transform into towns that transform into cities to medical breakthroughs that transform the way we treat diseases. We are surrounded by continuous change, so much that we often don't even think about it. But there are some transformations that are more acute than others and can't be ignored.

The AI transformation is one of these.

Whether you are excited about artificial intelligence or you are concerned by it, this is most certainly one of the most impactful transformations that the world has ever seen, and it will affect pretty much everyone in the modern world. The big question is: Is this going to prove to be a positive transformation, like a larva into a majestic butterfly, or a negative transformation from innocent to harmful?

I don't think anyone can perfectly predict the long-term outcome of AI's advance in civilization, but I do think that humans still have the ability to help direct this transformation. As an optimist, I'm excited about the AI transformation while at the same time, I am cautiously concerned.

The book you are reading does not intend to promote any particular agenda with AI. It does not propose any certain regulation, development, or direction for AI. Rather it simply intends to educate. The main goal is to show how AI is being used in various industries and in various settings, so people can make more educated decisions regarding AI practice and use. However, as the subtitle implies, this book also intends to help people thrive by embracing AI where it makes sense, so they can be one of those who takes advantage of the tools AI provides to reach their goals in work and life.

Since I'm not an AI expert, I didn't try to write this book about AI by myself. Rather, I solicited the help of thirty-three AI experts from various backgrounds and locations. The co-authors of this book come from all over the USA, Canada, Ireland, the Netherlands, Germany, Poland, Switzerland, Saudi Arabia, Brazil, and Japan. These authors have been featured on NBC, FOX, ABC, CBS, BBC, CSUN, and more. They are professionals who are university professors, scientists, business owners, consultants, product managers, engineers, software developers, cybersecurity leaders, healthcare experts, advisors, TEDx and keynote speakers, researchers, IT directors, CTOs, platform specialists, and more. The one thing these individuals have in common is that they all have something to share about artificial intelligence, and these ideas are available to you now.

Although this book is organized around the united theme of being prepared for the AI transformation that is sweeping across our planet, each of the chapters is totally stand-alone. The chapters in the book can be read in any order. I encourage you to look through the table of contents and begin wherever you want. However, I urge you to read all the chapters because, as a whole, they provide a great array of perspectives. Each is valuable in helping you understand how AI is being used in many areas of human industry.

I hope you discover ideas in these pages that help you successfully navigate civilization's next great transformation as humans, intelligent machines, and robots become increasingly interconnected and AI finds its way into more areas of everyday life. My hope is that this book encourages you to approach AI with curiosity rather than fear, giving you the confidence to embrace new opportunities and become a leader during this remarkable period of global change.

Whether artificial intelligence ultimately proves to be humanity's greatest opportunity or one of its greatest challenges, it is already transforming the world around us. The question is no longer whether AI will influence our lives, but how each of us will respond to that influence. My hope is that we choose to shape a future where humans and AI complement one another while continuing to value creativity and meaningful human relationships in ways that allow us to achieve our goals and live richer, more fulfilling lives.

Let's guide the AI transformation to help humanity avoid transforming into something harmful and rather sprout wings with AI to experience the world in a new and wonderful way.

About the Author

Erik Seversen is on a mission to inspire people. He holds a master's degree in anthropology and is a certified practitioner of neuro-linguistic programming. Erik draws from his years of teaching at the university level and years of real-life experience in business to motivate people to take action, creating extreme success in business and in life.

Erik is a TEDx and keynote speaker who has reached over one million people through his public speaking and live courses. He has visited over one hundred countries and all fifty states in the USA and has climbed the highest mountains on four continents, fifteen countries, and eighteen states. Erik has published twenty-four bestselling books on the topics of mindset, success, and peak performance, and he has helped over 500 people become authors. He is a full-time writer, book consultant, and speaker, and he lives by the idea that success is available to everyone—that living an extraordinary life is a choice.

Erik lives in Los Angeles with his wife and has two boys currently studying at university.

Contact Erik for interviews, speaking, or book publishing consultation.

Email: Erik@ErikSeversen.com

Website: www.ErikSeversen.com

LinkedIn: https://www.linkedin.com/in/erikseversen/

CHAPTER 1

INTELLIGENT REALITIES: KEEPING EDUCATION HUMAN IN THE AGE OF AI

By Amany Alkhayat, EdD
AI and Immersive Learning Leader
Queens, New York

We become what we behold.
We shape our tools and thereafter our tools shape us.
—Marshall McLuhan

In 1973, a man introduced as Dr. Myron Fox delivered an hour-long lecture to a room of psychiatrists, psychologists, and educators. He was warm, confident, articulate, and his audience rated him highly. Several found the talk stimulating. But there was no Dr. Fox. He was an actor who knew nothing about the subject, performing a script built entirely from jargon, contradictions, and nonsense. What fooled a room of experts wasn't the content. There was none. It was the performance of authority: the title, the confidence, the fluent delivery.

They mistook the feeling of being taught for the fact of having learned. Later researchers argued that the audience wasn't truly fooled, only charmed, but that only sharpens the warning: Either fluent authority manufactures false understanding, or we reward a smooth performance regardless of what's behind it.

Now imagine that the lecturer isn't a one-time hoax. Imagine they live in your students' pockets, available every hour, endlessly patient, always articulate, never short of a confident answer. That is roughly what an AI chatbot is: a voice fluent by design, with no idea whether it's right. And the gap the actor exploited in 1973, between sounding authoritative and being correct, is the one our students now walk into every time they ask a machine a question.

So, hold that lecture hall in your mind, because we are going to ask an uncomfortable question about your classroom. What if the greatest influence on learning is not intelligence, not the students', not the machine's, but the invisible authority of the realities we build them to learn inside? Students using AI without guidance feel independent while moving through answers already shaped, ranked, and delivered with confidence the content may not have earned. It is not a true agency. It is a plausible one.

Here is what I'll argue and the idea to hold onto throughout this chapter: An intelligent reality is any multimodal, AI-mediated environment—a chatbot on a screen, a virtual world, an augmented overlay—that actively shapes how learners perceive, interpret, and interact with knowledge. Notice what that doesn't say. Not "a smart AI." Not "VR goggles." The concept isn't about the machine's capabilities. It's about three things working at once: the reality a learner enters, the way it was designed, the algorithms that rank and filter, the persona an AI is given, the human choices baked into both, and the way the learner interacts with it. Perception is shaped in the space where those three meet. The environment is never neutral, because someone designed it, and that design quietly decides what feels true, what feels central, and who gets treated as the authority.

If the first claim is about reality, the second is about intelligence, and it's the one I most want to defend. The easy reading is that the machine is the intelligent one. I am arguing the reverse. The intelligence that decides an intelligent reality is human, the judgment to question a confident answer, to tell substance from performance, to keep what is worth keeping and discard the rest. AI can generate; only a person can lead. The danger was never that the machine is too smart. It's that we forget we are the smart ones and let the fluent voice take the wheel.

When educators hear "artificial intelligence," the conversation races to the machine: What can it do? Can it catch cheating? Can it tutor my students? Reasonable questions, but it's like asking how fast the car goes without asking where the road leads. And the machine is always happy to drive. The danger isn't that AI is a poor driver; it's that it steers smoothly while we ride along, mistaking motion for progress. The educator's job is to keep both hands on the wheel, aim at the outcomes that matter, and lead the technology there. So the better question isn't "How capable is it?" but "What kind of reality does it create, and does that road arrive at learning?"

There's a second question most education writing skips: Who built the voice at the lectern? When a student opens ChatGPT, Copilot, or Gemini, or talks to an AI inside an immersive environment, they enter a reality designed upstream by companies whose incentive is engagement, not your learning outcomes. Dr. Fox had a fake CV; today's confident voice has a business model. Educators don't author these realities; we choose among options someone else built and tuned. Our agency, too, is only a plausible one, unless we design with that fact in full view. If we are going to design these realities on purpose, the place to start is the problem that hit classrooms first and hit hardest.

Integrity Is a Design Problem, Not a Detection Problem

In late 2022, educators began noticing strange work: smooth, logical, typo-free, and somehow empty, like a beautifully built house with no one living in it. ChatGPT had arrived. Institutions reacted with policy and panic: prohibition clauses, detection software, faculty debates about surveillance.

Underneath was a fair fear: If a machine can produce the thing we grade, what have we been measuring? That question is uncomfortable, and it's the most productive one in education right now, because the essay was a flawed measure of learning before ChatGPT. A polished essay tells you a student could produce a polished essay by a deadline. It doesn't tell you whether they understand the material, can defend it, or connect it to something new. AI didn't break the assessment. It revealed cracks we'd ignored for decades. The educators who responded best didn't build taller walls. They redesigned the room.

Clarkson University stopped focusing on the final draft. Faculty in first-year writing knew that good writing rarely arrives fully formed; it emerges through false starts and revision, and the struggle is often where learning happens. So rather than banning AI, they changed the reality around it. Students wrote a rough draft entirely on their own, then uploaded it with the rubric and asked ChatGPT to critique it. The crucial move came next: Students couldn't just revise. They wrote a rebuttal note explaining which suggestions they accepted, which they rejected, and why.

The classroom conversation shifted. Students stopped chasing better essays and started debating the quality of the AI's advice: whether a revision strengthened an argument or just made it generic, whether it sharpened their voice or diluted it. They stopped treating AI as an authority and started treating it as one source of feedback among many. The learning had moved into the space between draft and revision, where students had to exercise judgment. The lesson: When assessments reward production, AI is a shortcut; when they reward reasoning, students are pulled back into the work. You can adapt this with four stages: draft, AI feedback, revision, reflection, and weight the reflection most heavily. That's where thinking becomes visible.

King's College London taught students to catch the machine lying. The faculty in psychology grew tired of the detection arms race and asked a better question: Even with perfect detection, students will graduate into workplaces full of AI. What they'll need isn't avoidance but evaluation. So students used Microsoft Copilot to turn a dense research article into a science blog post, then compared the AI's version against the original, line by line, as fact-checkers. They flagged where it oversimplified, dropped methodological caveats, or invented details about participants that never existed in the study.

The AI didn't write badly; it wrote convincingly, fluent enough to trust. But one student found the polished summary led to nearly the opposite of the researchers' point, and several said catching its errors was harder than writing the thing themselves. This is the fluency effect: We trust what sounds clear and confident, and AI is fluent by default, so learners mistake fluency for accuracy. To adapt it: Have students AI-summarize a hard reading, then grade their analysis of its failures, not the summary.

The University of South Australia made understanding harder to fake. Written exams were stressful and increasingly gameable, so bioscience faculty added interactive oral assessments that are short, structured ten- to fifteen-minute conversations where students explained concepts and handled follow-up questions. Preparation changed: Students no longer studied to reproduce text but to think aloud and defend reasoning. Understanding became visible, when students knew the material, it showed in how they handled the unexpected question; when they didn't, that showed too. Success was no longer "produce the right answer" but "show how you think." Start small: one low-stakes oral check-in or a short recorded explanation of one decision a student made in an assignment. Let students prepare topics, not exact questions.

Bansomdejchaopraya Rajabhat University in Thailand used AI as a brainstorm partner, not an author. Third-year English students stuck on the blank page were allowed to use ChatGPT, Gemini, and Bing, but only at the beginning, to generate angles. The AI-supported group produced more and more varied starting points. Yet faculty noticed the AI's ideas, however organized, lacked depth and emotional texture. The students still had to supply meaning. Think of it as skeleton and heartbeat: AI offers the skeleton that is the structure, options, and momentum; the student must provide the heartbeat, voice, interpretation, and lived context. Allow AI only in brainstorming, require students to explain which angle they chose and why, and grade the development of the idea, not its AI-generated seed.

A pattern emerges across all four: The most successful educators weren't the ones who used the most AI or the ones who fought hardest to keep it out. They were the ones who, with deliberate guidance, redesigned the work to live where humans still outperform the machine, judgment, voice, reasoning under question, so that doing well meant doing the thing AI can't.

So far in this chapter, AI has lived on a screen, an app on a phone, a chatbot in a browser, a tool you type at and read back. The next shift moves it off the screen and into a simulated world, where students don't write about a skill but step inside and perform it there.

When Intelligent Reality Becomes the Immersive Practice Space

Conversational AI is no longer entering education only through writing tools. Paired with extended reality, it shapes the spaces where students rehearse, fail, and build confidence: talking to virtual patients, responding to simulated students, and receiving real-time feedback. XR supplies spatial interaction; conversational AI supplies communication and meaning-making. Together they don't just add tools; they create new intelligent realities for learning. So the educator's question grows from "Did the student write this?" to "What kind of learning does this reality produce?"

You can read everything ever written about calming a frightened patient, memorize the script, recite the steps, and still freeze the first time one grips your arm and asks if they are going to die. That gap closes only one way: by living that exact moment, the grip, the question, the panic climbing your own chest, and getting it wrong with no one real on the other end, again and again, until the day you get it right. That is what AI-driven immersive simulations now make possible.

Picture a nursing student in an Oxford Medical Simulation "Hands and Voice" scenario. The patient is frightened and getting more so. The student is drawing up the medication, eyes on the syringe, hands steady, while talking the patient down, and the voice has to stay calm even as the hands do the precise, anxious work. The AI patient isn't following a script; it's reacting to the student's tone in real time, growing more agitated if the reassurance falters. The contribution of these simulations isn't repetition. It's presence. Situated in an environment real enough to trigger a genuine social response, the student doesn't recite lines into the air; their body reacts as though the patient were real. Decades of research on social presence point the same way: The more a virtual human feels like a real other, the more authentically we respond to it, which is precisely the demand a frightened patient or a disengaged teenager makes on us. For human skills, that realism is the whole point.

However, a vivid immersive world and a fluent chatbot can still add up to nothing learned. When the AI sticks to a script, the learner rehearses a smooth, tidy exchange that real people never actually have. In one language-learning study, learners called the AI's talk socially off: It clung to one topic, never interrupting or veering the way a real person does. So a student could ace the conversation inside the headset and still freeze when a real one refused to follow the script. The immersion was high; the transfer was not. What failed wasn't the realism but authenticity underneath, a conversation smoothed into something rehearsable teaches the wrong thing.

At one nursing school, 1,500 students needed to rehearse emergency hand-offs to a doctor, and there were only a few hundred medical students to play the part. So they built an AI doctor anyone could practice with, anytime. One student went from freezing up to knowing her lines cold; they'd run it twenty times. But the reps had a limit: students who practiced with real people walked away more confident in real conversations. The AI built the skill; people built the nerve.

So educators should use AI for the reps, not the relationship. Let it carry the early, repeatable practice, the part students normally get only once, so that scarce human time is spent where it actually counts: the real, unscripted conversation no machine can stand in for.

The greatest contribution of an immersive simulation, then, isn't how real it feels inside. It's whether it opens a door out, back to the real patient, the real classroom, the real conversation. So how do we bridge the gap between simply completing an AI-powered immersive simulation and actually learning from it, evaluating our decisions, uncovering hidden assumptions, and pulling out what transfers?

Intelligent Realities and the Role of Reflection

Practice alone doesn't produce learning. Students can finish a simulation and move on without understanding what happened. Reflection is where growth occurs, often after the activity, not during it. The simulation is the game; reflection is the film review that reveals the patterns. Afterward, ask three short questions: What did you learn about the skill? What did the simulation do well? What might be different with a real person? That last question matters most. It reminds students that the simulation is a representation, not reality, and that real people are always more complex, unpredictable, and emotionally nuanced than any model.

Does AI Build Thinkers or Dependents?

The technologies in these examples differed; the successful educators shared one habit. They stayed focused on what students would be capable of when the technology was gone. That gives us a test for any new tool: Will this help students become more capable of independent thinking or more dependent on the technology itself?

The question cuts through the marketing. A tool may be engaging, impressive, even time-saving. But if students become less able to reason, communicate, or judge without it, it isn't serving the educational goal; it's serving someone else's. And there's an access question underneath: "Just have students practice with AI" assumes reliable devices, paid tools, and steady bandwidth, so, unexamined, our most elegant solutions can widen the very gap this work exists to close. Designing on purpose means designing for the students who can't summon these tools at will.

The goal was never to produce students skilled at using technology. It's to produce people who can think, adapt, and contribute in a world where technology is everywhere. There will always be a confident voice at the lectern; our job is to make sure our students can tell the performance from the proof, listening past the fluency, questioning what stands behind it, and, when it matters, thinking past it to what was never actually said.

The technology in your classroom will keep changing. The students in your classroom will keep being human. Build for the humans.

About the Author

Dr. Amany Alkhayat is an AI and immersive learning leader, researcher, and UX designer who builds conversational AI tutoring systems and virtual and mixed-reality learning environments. She holds an EdD in instructional technology and media from Teachers College, Columbia University. She teaches AI courses at the Media University of Applied Sciences, has served as a mentor at Meta and for the NYIT Design Sprint, and has sat on the jury for The Polys XR awards. Previously, she taught VR and UX design, UX research, and human-computer interaction at the New York Institute of Technology. She has reached more than 7,000 students across three continents and delivered over 200 workshops on integrating AI and immersive technology into education. Her central argument is that purposeful, pedagogically grounded design is not optional; without it, AI produces tools that impress users without meaningfully supporting learning.

Website: https://www.dramanyalkhayat.com

LinkedIn: https://www.linkedin.com/in/amany-alkhayat/

CHAPTER 2

AI INCREASES INTELLIGENCE. IT DOES NOT INCREASE WISDOM

By Scott Beal
COO, PathO3Gen Solutions
St. Petersburg, Florida

Not everything that can be counted counts,
and not everything that counts can be counted.

—William Bruce Cameron

Years ago, while working in Intermodal Control at UPS, I was responsible for helping ensure that the company's Three Day Select product could move across the country without requiring additional transportation assets. This role demanded a level of operational discipline, as the network needed to absorb increasing volume while continuing to move packages efficiently from origin to destination using the resources already in place. Efficiency was not simply a goal. It was a requirement built into the structure of the system itself.

As peak season approached, a detailed operational plan was developed using the most advanced modeling tools available at the time. Forecasting systems projected how volume would move through the network and determined how aircraft lift should be distributed across destinations. The plan was built by highly intelligent professionals who relied on data, models, and years of analytical training to guide their decisions. Before the plan was finalized, several of us who worked inside the network daily were asked to review it, not because the models were insufficient, but because experience offered a different kind of visibility.

Those of us responsible for managing the flow of product had developed an understanding that came from watching the system operate under real conditions. We had seen how it behaved during peak periods, how certain destinations responded differently to demand, and how patterns emerged that were not always obvious within the data. While reviewing the plan, one detail stood out to me. The amount of aircraft lift scheduled for Portland did not seem sufficient based on what I had observed in prior peak seasons.

I can still remember the moment of hesitation before speaking up. The plan had been built carefully, and the individual presenting it was both capable and confident. There is a natural tendency in those moments to assume that the data must be right, especially when it is supported by sophisticated tools and thoughtful analysis. At the same time, there was a quiet conviction that came from having seen this play out before. It was not something I could fully quantify, but it was something I recognized.

That tension between what the data suggested and what experience revealed is something many leaders encounter, though it often goes unspoken. In fast-moving environments, there is pressure to align quickly, to trust the model, and to move forward. Choosing to question that path requires a willingness to stand on something less tangible but often more grounded.

From my earlier experience managing load planning for flights leaving the Louisville hub, I had developed familiarity with how specific destinations behaved under stress. Portland had a tendency to absorb more volume than expected, often in ways that did not align neatly with broader network trends. Based on that experience, I suggested that the plan might fall short in that specific location. The peak planner, who had built the model, was confident in the forecast. The data supported his conclusion, and the system performed well across most scenarios. We discussed it briefly and ultimately agreed to move forward, each holding our perspective.

When peak season arrived, the first significant surge in volume quickly revealed the outcome. Portland exceeded its planned capacity by the equivalent of an additional full aircraft. Shortly afterward, the planner came to my office and acknowledged what had happened, asking how I had anticipated the issue. The answer was not rooted in superior analysis or better data. It came from having lived the system long enough to recognize patterns that did not present themselves clearly within a model.

When Intelligence Meets Experience

That moment stayed with me because it revealed something deeper about how decisions are made. The planner was highly intelligent, well trained, and supported by sophisticated tools that performed well across most of the network. The model itself was not flawed in a broad sense. It captured relationships, trends, and patterns with impressive accuracy. However, it treated all destinations as if they behaved according to the same underlying assumptions, and that is where it fell short.

Portland did not follow a predictable progression during peak season. While most destinations scaled proportionally with overall volume increases, Portland responded differently, often experiencing disproportionate surges that could not be explained through linear modeling. The system assumed consistency where variability existed. What it could not capture was the nuance that comes from repeated exposure to real conditions. Patterns that appear irregular within data often become recognizable through experience, particularly when those patterns are influenced by factors that are difficult to quantify.

This distinction reinforced a lesson that has continued to surface throughout my career: There is no substitute for real world wisdom when making important decisions. Analytical tools expand our ability to process information and identify patterns, making us more informed and more capable. At the same time, they do not eliminate the need for judgment. They do not replace the understanding that develops from being inside a system long enough to recognize when something does not fit the expected model.

A World Where Intelligence Is No Longer Rare

The tools we relied on during that time were advanced for their era, yet they pale in comparison to the capabilities available today. Artificial intelligence can now process enormous data sets, identify subtle correlations, and generate recommendations at a speed that was once unimaginable. This shift represents more than technological progress. It changes how we think about intelligence itself.

For much of history, intelligence was constrained by access. Access to information, access to expertise, and access to the time required to analyze both created natural limits. Organizations competed based on their ability to gather and interpret knowledge more effectively than others. Artificial intelligence is rapidly removing those constraints, making analytical capability broadly accessible across industries.

Today, tasks that once required teams of specialists can be performed in moments. Insights that once took weeks to uncover can now be generated instantly. As intelligence becomes more available, its role begins to shift. The advantage no longer comes from having access to information. It comes from knowing how to interpret and apply it effectively. This is where the distinction between intelligence and wisdom becomes more meaningful.

The Illusion of Certainty

As analytical systems become more powerful, they also become more convincing. Artificial intelligence can produce forecasts, recommendations, and scenario analyses that appear precise, structured, and well supported. Because these outputs are delivered quickly and with a high degree of apparent confidence, it becomes easy to assume that they are correct.

There is also a human tendency at play that makes this dynamic even more powerful. People are naturally drawn to certainty, particularly in environments where decisions carry risk. When a system produces a clear answer, supported by data and presented with confidence, it offers something that feels reassuring. It reduces ambiguity and creates the impression of control.

Artificial intelligence strengthens that effect. The speed, clarity, and structure of its outputs can create a sense that the answer has been fully resolved, even when important variables remain unexamined. Without deliberate awareness, leaders can begin to substitute clarity for completeness, and confidence for accuracy.

This is where a subtle but important risk begins to emerge. Every model is built on assumptions, and every system depends on the data it receives and the patterns it has learned. Artificial intelligence does not eliminate these limitations. In many cases, it can make them less visible by producing outputs that appear complete and authoritative.

The Portland example reflects this clearly. The model performed well across most of the network and was effective within its defined parameters. What it could not account for was a specific behavior that fell outside of its assumptions. Situations like this are not rare. Models tend to perform best when systems behave consistently and become less dependable when variability, human behavior, or external influences introduce complexity.

As artificial intelligence continues to evolve, the risk is not that it will produce poor analysis. The risk is that it will produce analysis that appears so complete that we stop questioning it. Wisdom requires a different posture. It requires the ability to engage with analytical outputs while maintaining awareness of their limitations and asking what may not be captured within the model.

Leadership at the Human Edge

Artificial intelligence will continue to evolve and become increasingly integrated into how organizations operate. Its influence on decision making will grow, and its capabilities will continue to expand. At the same time, the fundamental nature of leadership will not change.

The responsibility that comes with this shift is significant. As intelligence becomes more accessible, the weight of decision making does not decrease. It increases. Leaders are no longer limited by a lack of information. They are challenged by an abundance of it and by the need to determine what truly matters within it.

In that environment, wisdom is not a passive trait. It is an active discipline. It requires reflection, restraint, and the ability to remain grounded when everything around you is accelerating. The future will not be defined by those who simply move the fastest. It will be shaped by those who understand when to pause, when to question, and when to lead with clarity in moments where certainty is only an illusion.

About the Author

Scott Beal is a senior business operator and current COO of PathO3Gen Solutions, where he focuses on scaling advanced technology across healthcare, life sciences, and clean manufacturing environments.

Scott began his career at UPS, developing deep operational expertise in load planning and industrial engineering within complex global logistics networks. He later transitioned into leadership roles centered on commercialization, growth strategy, and operational execution. He brings a practical perspective to artificial intelligence and leadership, grounded in real-world environments where decisions carry meaningful operational and human consequences, and he is also the creator of Scotty Velvet Media, a platform focused on thoughtful leadership, storytelling, and reflection in a rapidly evolving world.

LinkedIn: https://www.linkedin.com/in/scott-beal-77180b3/

CHAPTER 3

YOUR THINKING IS THE MOAT: HOW SMALL BUSINESSES THRIVE WITH AI

By Tanya J. Beck
Founder of Beck Strategic Advisory, AI Trainer
Birmingham, Alabama

Clear thinking becomes your competitive advantage.
—Shane Parrish

A Note Before We Begin: At the time of this writing, AI is moving faster than anything I've seen in my career. The frontier labs keep releasing models more capable than the last, and the distance between "impossible" and "ordinary" now closes in months, not years. The intelligence available to a small business owner today, for the price of a monthly subscription, would have read as science fiction a few years ago. Some of the specifics in these pages may age. By the time you read this, the models will almost certainly have moved past it.

That isn't what this chapter is about. It's about the thinking underneath, the ideas that hold, no matter which model you're using or how good it gets. That's the bet of this chapter, and it's the same bet I'd make for your business.

The Real Barrier Isn't the Technology

If you're reading this book, you've already cleared the first hurdle. You're not asking whether AI matters. You've decided it does, or at least that it matters enough to spend an evening reading about it. I'll skip the part where I convince you and go straight to the part nobody talks about: what this actually looks like in the day-to-day work of running your business.

That's where a lot of small business owners get stuck. Not on fear, and not on doubt. On a quieter thing. They've been told, loudly and from every direction, that AI changes everything. What they haven't been told is where to put their hands. The hype handed them urgency without a single handhold. That gap, between "this matters" and "I don't know where to start," is the real barrier. Not the technology.

It doesn't help that the loudest voices live at the extremes. One side promises a revolution by next quarter. The other warns it's all going to break. Neither is much use when you've got nine employees and a problem you've been meaning to fix for a year. I've always liked the way a friend of mine, Austin Sensman, who founded Birmingham AI and Caravan AI, puts it: "AI is not happening to you. It's happening for you." I think that's exactly right. It isn't a force of nature you need to brace for, and it isn't a magic box that runs your business while you sleep. It's closer to a new kind of colleague. Fast, capable, occasionally brilliant, and not always right. Something you direct and then check the way you'd direct and check anyone new. That single shift, from the thing that happens to me to the thing I direct, is where everything starts.

Here's the part that's sometimes easy to miss: The place to point it first isn't the big, impressive transformation. It's the small, tedious thing in the back of your mind. The task that eats a few hours every week. The one you do on autopilot while knowing there has to be a better way. There usually is. But finding it depends less on the tool than on a skill that has nothing to do with technology. The ability to step back and think clearly about your own work, before you ask anything, human or machine, to help with it. There's a name for that skill, and the rest of this chapter is about building it.

Capable and Brand-New

If AI is a new kind of colleague, it helps to be specific about what kind. Picture the most capable beginner you can imagine. Quick, tireless, genuinely sharp. It reads faster than anyone on your team, drafts in minutes, and never gets bored. And out of the box, it knows nothing about your business. Not your customers, not your standards, not the reasons behind how you do things.

That's what you start with. Enormous capability, no context. That leaves you two jobs, and they frame everything AI does for you. The first is to give it what it's missing. The second is to check its work, because capability without your judgment isn't worth much. The first job gets easier over time, and that's the good news. Good AI tools let you teach them about your business, and they remember. You're not starting over in every conversation. The more clearly you explain how you work, the sharper the help gets, so the time you spend describing your business isn't overhead. It builds on itself. The second job never goes away, and it's the part to remember. No matter how well you've taught it, AI will not tell you when it's unsure. A real new hire raises a hand and says, "I don't know this one." AI doesn't. It hands you a wrong answer with the same confidence as a right one. That single trait is why a person stays in the loop. Not because the tool is weak, but because confident and correct are not the same thing, and on the things that matter to your business, only you can tell the difference. Teach it, and check it. The tool brings the speed, and increasingly the memory. You bring the judgment. Hold on to that, because the rest of this chapter is three ways to practice it.

Finding the First Problem

The hardest part of starting with AI usually isn't the tool. AI is turning up inside the software you already use, in apps built for one job, in assistants that will attempt almost anything. That flood is exactly why chasing tools is the wrong move. The question was never which tool. It's where to point one. Most businesses go looking for a place to use AI, which reverses the natural order. You don't hire help and then invent a job for it. You start with the work that needs doing.

Start by following the friction. Somewhere in your week is a task that quietly drags, and it tends to have three markers. It repeats, so you do it on a schedule, not once. It drains more time than thought, so it's labor, not real judgment. And you'd catch a mistake instantly, because you know the work well enough to know when it's wrong. A task with all three is the back-of-mind chore you've been meaning to fix for a year. It's also the safest, smartest place to begin, because the payoff repeats and the risk is low. Once you've found it, ask three questions before you build anything. Answering them honestly might take an afternoon. Skipping them can cost you two months.

The first question: What problem is this actually solving? Name it in one plain sentence. Not "I want to use AI," but "I spend two hours every Monday turning the same raw numbers into the same report." If it takes more than a sentence to explain, there's a good chance the core problem is still hiding underneath, and that's worth knowing before you spend a dollar.

The second question: What's the cost of doing nothing? Every routine task has a price, even when it never shows up on an invoice. Hours lost, errors repeated, follow-ups that slip. Some problems turn out to be expensive enough to fix this week. Others, once you look squarely at them, aren't worth the trouble yet. Both answers are useful. The point isn't to justify using AI. It's to find out which problem actually deserves to go first.

The third question: What does "solved" look like? This is the easiest step to overlook, but it matters most. "Better reporting" is a wish. "The Monday report builds itself and lands in my inbox by

8 a.m." is a finish line. Describe the finish line in plain words and something remarkably powerful happens. You can now tell whether AI is even the right tool. Sometimes it is. Sometimes a smarter spreadsheet would do, and you've just saved yourself from building a complex solution you didn't actually need.

To see how this works in practice, let's look at a commercial equipment maintenance company. The dispatcher was consistently receiving urgent emails from busy facility managers reporting machine issues. Because every client had their own way of writing these requests, the dispatcher spent an hour or two every day carefully reviewing the messages to locate the specific machine model, the location on site, and the error code needed to create a standard dispatch ticket for their technicians. It hit all three markers of friction.

Before looking for an AI tool, they asked the three questions: (1) What problem is this actually solving? They named it in one plain sentence: "I spend an hour a day manually converting differently formatted client requests into our standard dispatch tickets."

(2) What's the cost of doing nothing? The cost here wasn't just the lost hour. The real price was the occasional "follow-up that slipped": a technician arriving without the full context of the machine's error codes, leading to lost billable hours while they pieced together the issue on site.

(3) What does "solved" look like? They defined the finish line: "Incoming client requests are automatically reorganized into our standard ticket format, with the machine model, address, and core issue clearly separated so I can simply review the information, click 'approve,' and send the tech."

None of this is technical. Notice the friction, name the problem, define what done looks like. That's the thinking that decides whether AI ever earns its keep.

the three foundational questions:

1 **What problem is this actually solving?**
They named it in one plain sentence: "I spend an hour a day manually converting differently formatted customer requests into our standard dispatch tickets."

2 **What's the cost of doing nothing?**
Just as every routine task has a hidden price, the cost here wasn't just the lost hour. The real price was the occasional "follow-up that slipped", a technician arriving without the full context of the machine's error codes, leading to lost billable hours while they pieced together the issue on-site.

3 **What does "solved" look like?**
They defined the finish line: "Incoming client requests are automatically reorganized into our standard ticket format, with the machine model, address, and core issue clearly separated so I can simply review the information, click 'approve,' and send the tech.

Knowing Good from Bad

Once AI is doing real work for you, the job that's left is the one that matters most, and it isn't technical. It's judgment. Someone has to look at what came back and know whether it's any good.

That sounds obvious, but the tool's polished output makes it surprisingly easy to overlook. Well organized, confidently worded, formatted like something a professional would hand you. The polish is what makes it tricky, because polish is not the same as accuracy. AI is very good at sounding right. Telling whether it actually is right is your part, and it will stay your part no matter how capable the tool becomes.

Here's the reassuring thing for a small business owner. You are already equipped for this. You know your customers. You know what your best work looks like and what falls short of your standards. You know the number that's plausible and the one that can't possibly be right. That hard-won knowledge of your own business is exactly what AI lacks and exactly what good judgment requires. You're not unqualified to check the machine. You're the most qualified person in the building. An unexpected value emerges when you do it. Every time you look at a draft and decide what's good, what's off, and what's missing, you're catching the errors and sharpening your own sense of what "good" even means. The standard gets clearer in your head, which makes your next instruction better, which makes the next draft better. Judging the work and improving the work turn out to be the same loop.

The rule is simple: Let AI produce, and keep the judging for yourself, especially on anything a customer will see or a decision will rest on. Start where you can check the work at a glance, which is exactly why "you'd catch a mistake instantly" made the list of a good first problem. As your trust grows, you can ease up on the low-stakes things. But the judgment itself, the knowing good from bad, is not something you'll ever want to hand away. It's quietly becoming the most valuable thing you do.

Start Small, Then Spread It

You don't have to transform anything. The businesses that get real value from AI almost never started with a grand plan. They started with one frustrating problem, solved it, and let that win suggest the next one.

It helps to move in that order on purpose. First, solve the one problem you found by following the friction. Build the smallest version that actually works, with your judgment checkpoint built in, and run it on real work until you trust it. That's the whole first lap. If it feels much simpler than you anticipated, you are doing it right. It's supposed to teach you what this looks like inside your business, with low stakes and a short distance to fall.

Then, and only then, expand. Point what you learned at the next problem. The second build is faster because you know what to look for. The third is faster still. That's when the compounding starts, and it only happens after you've laid this foundation and built that initial trust. The owners who move fastest over a year are usually the ones who moved deliberately in the first month.

Here's the step that separates early exploration from genuine adoption. At some point, AI has to stop living inside one person's head.

In a lot of small businesses, AI is one curious person. The owner who reads about it on weekends. The one team member who pays for a subscription out of pocket and keeps bringing it up. That's not adoption. It's a pilot light.

The shift comes when someone decides to bring the whole team in, not just license a tool. This is where a lot of businesses quietly stall. They buy the access, hand it around, and stop there. But access isn't skill. What changes things is training, and not a one-hour lunch-and-learn, but real time with the actual work. When I train corporate teams, someone always walks in, certain this isn't for them because they don't write code. Then something shifts. They build one small thing themselves and realize their existing business expertise is the exact skill needed to turn a piece of software into real business value. A month later they've automated something I never taught them, because once people see what's possible, they seek out their own problems to solve. A few things separate the teams that thrive from the ones that stall. Someone owns AI as a named responsibility, so it isn't everyone's job and, therefore, no one's. There's a short written policy, what data is off-limits and when a human has to check the work, so the careful stop self-censoring and the over-eager stop oversharing. Leadership pays for real usage, not just access, because buying a seat is cheap and the compounding value only shows up once people use the tools enough to find it. And people have genuine permission to experiment, with the time and the blessing to try things, not just on their own after hours.

That's the arc. One problem, then a pattern, then a team that owns it. Start small enough to succeed, and spread it before it gets stuck in a single pair of hands.

Your Thinking Is the Moat

Here's the shift underneath all of this, the one that anchors everything else. AI makes execution almost effortless. The tasks that used to take real time and skill, the drafting, the sorting, the first version of almost anything, are becoming fast and nearly free. That's the part that unsettles people, and for a small business it can feel especially pointed. If the basic drafting and sorting become commoditized, what is your true competitive advantage? The answer is the part that was always actually yours. Not the doing, but the deciding. Knowing which problem is worth solving. Knowing what "good" looks like for you. Knowing when the confident answer in front of you is wrong. While execution matters, your unique edge has always been your judgment. AI doesn't take that. It raises its value, because when the doing gets cheap, the deciding is what's scarce.

That's the skill this whole chapter has been about, and it has a name: metacognition. The ability to think clearly about your own thinking. The people who get the most out of AI aren't necessarily the ones with the deepest technical backgrounds; they are the ones who can build clear mental models of their work, break a problem down into simple steps, articulate exactly what they need, and clearly state what "solved" looks like. Noticing the friction, framing the real problem, judging the result. It's worth naming because it's the one thing no tool will ever make obsolete. The technology underneath all of this will keep changing, faster than any book can track. Every leap forward makes the human element the most critical factor. The person who has to decide what's worth doing and say what good looks like. The better the tools get, the more your thinking is the thing that sets your business apart.

You don't have to be ahead of the curve. You don't need the biggest budget or the most technical team. You just have to think clearly about your own business and pick the first problem worth solving.

Solve the right problem first. The technology will follow.

About the Author

With nearly thirty years of experience in corporate marketing, Tanya Beck has built a career grounded in clear strategy and a relentless curiosity that constantly drives her to find a better way. Her approach was shaped by working alongside talented people and learning from leaders who deeply influenced how she thinks about execution. Today, she runs Beck Strategic Advisory, using practical technology to solve everyday problems for small businesses. As a training guide for Caravan AI and a partner with Kinetic Communications, she teaches teams how to smoothly integrate the right tools into their actual workflows. Born in London, raised in Hong Kong, and based in the US for over three decades, Tanya brings a uniquely global perspective to her work.

Email: tanya@beckstrategicadvisory.com

Website: www.beckstrategicadvisory.com

LinkedIn: www.linkedin.com/in/tanya-j-beck/

CHAPTER 4

COMMANDING RESILIENCE: AI'S ROLE IN ADVANCING CRISIS LEADERSHIP AND PUBLIC SAFETY SYSTEMS

By Dr. De'Anjelo J.L. Bradley
Global Executive Leader, Author of *The Antithesis*
Atlanta, Georgia

Crisis leadership today requires the ability to navigate complexity, manage large streams of information, and make rapid, critical decisions. Unlike traditional leadership, which prioritized command presence, personal experience, and strict adherence to established rules, contemporary crisis leadership demands greater adaptability and readiness to respond to uncertain, continually evolving circumstances.

Modern public safety leaders operate within interconnected, unpredictable, and data-rich environments that challenge conventional approaches. In this evolving landscape, artificial intelligence commands a pivotal role by transforming how leaders interpret information and respond to crises.

Traditionally, resilience meant the ability to recover from disruptions. In public safety today, resilience also involves adapting, evolving, and maintaining operations during ongoing stress. AI supports this modern view by providing predictive insights, accelerating decision-making, and enabling systems to adjust. AI does not replace leaders; it gives them better tools for exercising command in uncertain situations.

This chapter introduces "commanding resilience": a leadership model that combines human judgment with AI to improve decision-making, awareness, and adaptability. By connecting this model to the new idea of resilience, the chapter shows why combining human expertise with machine intelligence is vital for effective crisis leadership and better public safety outcomes.

Traditional emergency management systems, including the Incident Command System (ICS) and the National Incident Management System (NIMS), were designed to establish order within chaos.[1]

These systems focus on clear chains of command, well-defined jobs, and consistent communication methods. While these are still important, they were made when information was scarce, slow to process, and mostly focused on what had happened.

Today's operational environment compels a shift from reactive to anticipatory crisis leadership. Leaders must not only manage incidents as they emerge but are now required to preemptively identify and address risks before escalation. AI operationalizes this imperative, converting both historical and live data into actionable forward-looking intelligence.

With machine learning, predictive analytics helps leaders identify trends and patterns that were missed in older methods.[2,3] For example, AI can predict when infrastructure might fail by analyzing repair records, environmental factors, and usage patterns. This lets public safety leaders act early and lower the chance of serious problems.

This change has created a new type of crisis leadership. Commanding resilience means leaders focus on the present and the future. For example, during a health crisis, artificial intelligence can predict how a disease spreads, helping crisis leaders dispatch medical help early and act before the situation escalates. With these tools, leaders prepare ahead and respond quickly.

Crisis leadership fundamentally demands making rapid, high-stakes decisions. Decisions rendered during crises often determine outcomes of life or death. AI equips leaders with superior decision-making capabilities by providing immediate and comprehensive information beyond human reach.

AI can process huge amounts of information at once, finding links and patterns people might not see. This includes sensor data, messages, photos, and social media.[4]

Natural language processing (NLP) enables AI to extract useful information from emergency calls, reports, and public messages.[5,6,7]

Using AI in decision-making fits with the OODA loop—observe, orient, decide, act—from military strategy.[8] AI speeds up observation and understanding by quickly sorting data, enabling leaders to make decisions and act faster. This is vital when time matters.

AI is an indispensable asset that amplifies human capabilities while never replacing them. Leaders retain ultimate authority and responsibility, drawing on AI-informed analysis to achieve the highest standards for crisis leadership.

Situational awareness is imperative for leaders in crisis. Endsley (1995) defines situational awareness as the ability to perceive current events, interpret their significance, and project potential developments.[9] AI augments each of these stages, transforming static awareness into a dynamic, evolving operational perspective. AI pulls data from many sources, such as smart devices, drones, satellites, and social media. Putting it all together gives a clear view of what is going on.

During disasters, for example, AI can collect weather, road, and building conditions, as well as crowd movement, to produce real-time maps. AI does more than gather data. It helps leaders by identifying patterns and anomalies. Machine learning can spot anomalies and warn leaders early about potential problems. This is especially important in large systems, where early warnings help leaders act quickly.

Projection is the highest level of awareness, where AI's predictive power is most helpful. By running simulations of potential outcomes, AI lets leaders see possible futures and adjust plans before problems get worse. This is similar to digital twin and advanced modeling.[10] Looking ahead like this is key to commanding resilience in complex situations.

AI-Enabled Operational Resilience

Operational resilience means public safety systems can continue working and recover quickly after incidents.[11] Artificial intelligence enhances operational resilience by helping systems adapt, maintain backups, and continuously learn.

Artificial intelligence's main strength is learning. As machine learning processes more data, it improves, helping public safety systems adapt as conditions change.[12,13] This means every crisis event makes the system more resilient for the future. AI also helps use resources better. In crises, it's key to get people, tools, and supplies where they are most needed.

AI models consider need, what's available, and logistics to identify the best locations for resources.[14] This way, resources go where needed and are used wisely. AI also helps keep systems running by spotting problems early and fixing them sooner. For instance, in transport, AI can reroute traffic after crashes or jams to keep things moving and cut risks.

Training and simulation represent crucial domains in which AI fortifies resilience. AI-driven simulations generate realistic, challenging scenarios that compel leaders to hone their decision-making in complex operational environments. Mahdi et al. (2023) confirm these exercises enhance real-world outcomes by increasing leaders' preparedness, decision speed, and adaptability.[15] AI-enabled training decisively builds confidence and critical capabilities, ensuring leaders excel during actual crisis response.

Ethical Leadership in AI-Driven Systems

Bringing AI into public safety systems raises important ethical issues. For example, law enforcement agencies' use of facial recognition technology can lead to the misidentification of minority individuals, causing wrongful detentions and reinforcing bias. Predictive policing algorithms have also been criticized for unfairly targeting certain communities due to biased historical data, which continues existing inequalities. Another concern is AI-driven risk assessment tools in the justice system, which sometimes make biased suggestions about bail or sentencing. Leaders need to address data privacy, algorithmic bias, and accountability to ensure AI is both fair and effective.[16]

A big concern is algorithmic bias. AI is only as fair as the data it learns from. If old data is biased, the AI may replicate or exacerbate those biases.[17,18] Leaders are responsible for instituting regular evaluations and ongoing adjustments to AI systems to actively minimize bias and advance fairness.

Data privacy is a further crucial issue. AI systems often rely on large datasets that may include sensitive personal information. Leaders must manage the need for analytics-based insights with the protection of individual rights, while adhering to legal and ethical standards.[19]

Accountability is still a central principle. While AI can provide recommendations, the responsibility for decisions ultimately lies with human leaders. This underscores the importance of maintaining human oversight and ensuring that AI systems are transparent and explainable.[20,21]

Ethical leadership in the context of AI necessitates a commitment to integrity, transparency, and accountability. It also demands continuous education and awareness to ensure that leaders comprehend both the capabilities and limitations of AI technologies.

The Human Dimension of Command

Despite the transformative potential of AI, the human dimension of leadership remains essential. Crisis leadership extends beyond data and analytics to encompass communication, trust, and the ability to motivate and lead individuals in times of uncertainty.

AI cannot duplicate emotional intelligence, empathy, or the ability to build relationships. These human qualities are indispensable for effective leadership, particularly in high-pressure environments where morale and cohesion are critical.

AI can augment the human dimension by reducing cognitive load. By automating routine tasks and providing decision support, AI enables leaders to focus on the strategic and interpersonal aspects of leadership. This facilitates more effective engagement with teams and stakeholders.[22]

The infusion of AI into leadership is best conceptualized as a partnership. Human leaders contribute judgment, contextual understanding, and ethical guidance, while AI offers analytics-based insights and analytical functions. This collaboration results in a more effective and resilient leadership model.[23]

Future Trajectory: AI and the Next Generation of Crisis Leadership

The role of AI in public safety systems is anticipated to expand substantially. Emerging technologies, including digital twins, autonomous systems, and advanced predictive analytics, are projected to further augment the capabilities of crisis leaders.[24]

Digital twins, for example, create virtual representations of physical systems, enabling real-time simulation and optimization. In airports and cityscapes, digital twins can model operations and predict the impact of disruptions, supporting more effective planning and response.[25]

Autonomous systems, including drones and robotics, can operate in high-risk environments, reducing risk to human responders.[26] AI-driven decision systems will continue to evolve, providing increasingly sophisticated analysis and recommendations.

This evolution necessitates continuous adaptation among leaders. AI literacy is poised to become a core competency, and organizations must invest in training and development to ensure leaders are equipped to utilize these technologies effectively.[27] Specific organizational training approaches may include structured AI literacy programs, scenario-based simulations, ongoing workshops, and collaborative exercises that focus on both technical skills and the ethical use of AI systems.[28]

At the organizational level, integrating AI requires a strategic approach. This includes investment in infrastructure, data management, and cybersecurity, as well as a commitment to innovation and continuous improvement.[29]

Conclusion

Commanding resilience in the contemporary era necessitates a redefinition of leadership. This redefinition involves integrating human judgment with machine intelligence to navigate increasingly complex and dynamic environments. AI offers a robust mechanism for enhancing decision-making, situational awareness, and operational resilience, thereby enabling leaders to achieve superior performance.

The lasting impact of AI on crisis leadership will depend on its successful integration with leadership practices, guided by both technical acumen and principled judgment. To facilitate this synthesis, leadership development programs must evolve to foster not only AI literacy and technical skills but also robust ethical frameworks that prioritize transparency, fairness, and human accountability. Systematic education on the responsible use of AI will enable leaders to leverage these tools in ways that amplify, rather than diminish, essential human capacities such as empathy, discernment, and moral reasoning. Organizations should invest in dynamic training strategies, including scenario-based simulations, technical workshops, and continuous professional development to prepare leaders for the nuanced realities of AI-enabled crisis environments. Looking forward, the next generation of crisis leaders will be defined by their ability to navigate the intersection of advanced technologies and human-centered leadership—achieving resilience and adaptability through a thoughtful balance of innovation and enduring human values.

As public safety systems evolve, successful leaders will be those who can command resilience. These leaders will recognize AI's potential, embrace its capabilities, and integrate it into their leadership strategies. By doing so, they will not only respond to crises more effectively but also contribute to the development of stronger, more adaptive, and better-prepared systems.

References

1. Federal Emergency Management Agency, *National Incident Management System*, 3rd ed. (Washington, DC: U.S. Department of Homeland Security, October 2017).
2. Ian Goodfellow, Yoshua Bengio, and Aaron Courville, *Deep Learning* (Cambridge, MA: MIT Press, 2016).
3. Christos Tsallis et al., "Application-Wise Review of Machine Learning-Based Predictive Maintenance: Trends, Challenges, and Future Directions," *Applied Sciences* 15, no. 9 (2025): 4898, https://doi.org/10.3390/app15094898.
4. Goodfellow, Bengio, and Courville, *Deep Learning*.
5. Kevin Gormley, Katy Lockhart, and Jolly Isaac, "Using Natural Language Processing in Facilitating Pre-hospital Telephone Triage of Emergency Calls," *British Paramedic Journal* 7, no. 2 (2022): 31–37, https://doi.org/10.29045/14784726.2022.09.7.2.31.

6. Jonathon Stewart et al., "Applications of Natural Language Processing at Emergency Department Triage: A Narrative Review," *PLOS ONE* 18, no. 12 (2023): e0279953, https://doi.org/10.1371/journal.pone.0279953.

7. Hao Wang et al., "Using Natural Language Processing in Emergency Medicine Health Service Research: A Systematic Review and Meta-analysis," *Academic Emergency Medicine* 31, no. 7 (2024): 696–706, https://doi.org/10.1111/acem.14937.

8. Varun Sehgal, "Artificial Intelligence and Military Decision Making: Revisiting OODA Loop Framework," *International Journal for Multidisciplinary Research* 6, no. 4 (July–August 2024): 1–6, https://doi.org/10.36948/ijfmr.2024.v06i04.26389.

9. Mica R. Endsley, "Toward a Theory of Situation Awareness in Dynamic Systems," *Human Factors* 37, no. 1 (1995): 32–64, https://doi.org/10.1518/001872095779049543.

10. National Academies of Sciences, Engineering, and Medicine, *Foundational Research Gaps and Future Directions for Digital Twins* (Washington, DC: National Academies Press, 2024), https://doi.org/10.17226/26894.

11. National Institute of Standards and Technology, "Operational Resilience," *Computer Security Resource Center Glossary*, accessed June 27, 2026, https://csrc.nist.gov/glossary/term/operational_resilience.

12. Goodfellow, Bengio, and Courville, *Deep Learning*.

13. Tsallis et al., "Application-Wise Review," 4898.

14. Xiao-Bing Hu et al., "A New Method for Resource Allocation Optimization in Disaster Reduction and Risk Governance," *International Journal of Disaster Risk Science* 7, no. 2 (2016): 138–50, https://doi.org/10.1007/s13753-016-0089-2.

15. Syed Sarosh Mahdi et al., "Systematic Review on the Current State of Disaster Preparation Simulation Exercises (SimEx)," *BMC Emergency Medicine* 23, no. 1 (2023): 52, https://doi.org/10.1186/s12873-023-00824-8.

16. Elham Tabassi, *Artificial Intelligence Risk Management Framework (AI RMF 1.0)*, NIST AI 100-1 (Gaithersburg, MD: National Institute of Standards and Technology, 2023), https://doi.org/10.6028/NIST.AI.100-1.

17. Cathy O'Neil, *Weapons of Math Destruction: How Big Data Increases Inequality and Threatens Democracy* (New York: Crown, 2016).

18. Reva Schwartz et al., *Towards a Standard for Identifying and Managing Bias in Artificial Intelligence*, NIST Special Publication 1270 (Gaithersburg, MD: National Institute of Standards and Technology, 2022), https://doi.org/10.6028/NIST.SP.1270.

19. Tabassi, *Artificial Intelligence Risk Management Framework.*

20. P. Jonathon Phillips et al., *Four Principles of Explainable Artificial Intelligence*, NISTIR 8312 (Gaithersburg, MD: National Institute of Standards and Technology, 2021), https://doi.org/10.6028/NIST.IR.8312.

21. Tabassi, *Artificial Intelligence Risk Management Framework.*

22. Isangadighi Wisdom, "AI-Augmented Leadership: Enhancing Human Decision-Making," *International Journal of Advances in Engineering and Management* 6, no. 12 (December 2024): 582–85, https://doi.org/10.35629/5252-0612582585.

23. Chieh Peng Lin and Pei Chun Chen, "Mentoring for Effective Human-AI Collaboration: An Integrated Theoretical Framework," *Total Quality Management & Business Excellence* 36, nos. 9–10 (2025): 1062–76, https://doi.org/10.1080/14783363.2025.2504603.

24. OECD, "AI in Law Enforcement and Disaster Risk Management," in *Governing with Artificial Intelligence: The State of Play and Way Forward in Core Government Functions* (Paris: OECD Publishing, 2025), https://doi.org/10.1787/795de142-en.

25. National Academies of Sciences, Engineering, and Medicine, *Foundational Research Gaps.*

26. Thomas Manzini et al., "Harnessing AI and Robotics in Humanitarian Assistance and Disaster Response," *Science Robotics* 8, no. 80 (2023): eadj2767, https://doi.org/10.1126/scirobotics.adj2767.

27. Tabassi, *Artificial Intelligence Risk Management Framework.*

28. Thomas D. Wilson, "The Development of Policies on Generative Artificial Intelligence in UK Universities," *IFLA Journal* 51, no. 3 (October 2025): 722–34, https://doi.org/10.1177/03400352251333796.

29. Tabassi, *Artificial Intelligence Risk Management Framework.*

About the Author

Dr. De'Anjelo Bradley, PhD, is a leader in emergency management, a published author, and a global humanitarian committed to advancing safety, resilience, and innovation. With a distinguished career spanning military service, international fire leadership, and executive-level emergency management, he brings a rare blend of operational experience and strategic insight.

De'Anjelo has served in high-risk environments (Iraq, Afghanistan, Kuwait, Ecuador) across multiple countries, leading complex operations and shaping safety programs that impact millions. He is currently a senior leader in emergency management, where he oversees large-scale safety initiatives, training programs, and interagency coordination.

Beyond his professional work, he is the founder of multiple organizations focused on mentorship, veteran advocacy, and community development. As an author and thought leader, his work centers on resilience, leadership under pressure, and preparing individuals and organizations for the evolving challenges of the modern world.

LinkedIn: https://www.linkedin.com/in/dr-deanjelo-b-71b2b91b/

CHAPTER 5

AI: THE REALITY BEYOND THE FANTASY

By William Brendel, PhD
Engineering & ML Executive, Founder, AI Strategist
Reno, Nevada

The greatest danger in times of turbulence is not the turbulence. It is to act with yesterday's logic.

—Peter Drucker

A Groundhog Day
5:58 a.m.

"Darn it!" Your alarm is set for 6:15 a.m. But a notification pulses. You're not even awake yet, but you mechanically read it. A VC on LinkedIn declaring: "AI will eliminate half of all white-collar jobs by 2027." Below it, a hundred people arguing about whether he is being reckless or conservative. A hundred...you feel behind.

You put the phone down. You pick it up again. There are four more notifications, each one a different frequency of the same ambient alarm. You cannot even tell anymore which of them were written by a person. In a quiet home, chaos has invaded your mind. You spend the remaining ten minutes collecting your thoughts.

That last part is new. A year ago you could feel the difference between a human message and a generated one. Now the spam reads like your colleagues and your colleagues sometimes read like spam. Something that used to function as a basic social filter has stopped working. You have not found a replacement for it. Its disappearance feels like a loss beyond yourself.

6:14:51 a.m.

You have less than ten seconds to disable your alarm right before it rings. No need to add more noise to the disturbance.

Without even thinking, you go back to your socials. More comments. It feels like everyone is at war. You read four of them before you notice you are holding your breath. Your body screams to let go, but your mind is trapped in the game. The scroll is real. A competitor announced an AI-first transformation. You check your mailbox in a vain attempt to escape the madness and refocus on work. A newsletter wants to know why your industry is falling behind. An email from a board member, sent at 11 p.m., with a subject line that is just a question mark and a link to an article about automation. You give up, but still smirk: At least your shower will be phone-less.

7:38 a.m.

On your way to the office. Finally, some time to yourself to get mentally prepared for the coming day. The traffic has suddenly stopped. The self-driving car ahead of you is the reason. It drifts into the merge lane at precisely the speed limit, executing a maneuver that is perfectly legal but would never occur to a human driver paying attention to those behind them. The line of cars stacks up. No one honks because there is no point. Instead, you all film the scene. Everyone thinks, "Maybe it will go viral, and I can capitalize on it." The car does not feel social pressure. It feels like a distant scene from a sci-fi movie you used to watch as a kid. And now you're living it.

Four minutes pass. You watch the vehicle make a series of correct decisions that somehow produce a wrong outcome for everyone around it, and you think, briefly, that this is a decent metaphor for something. Finally, the lane opens and you drive.

8:31 a.m.

Your assistant meets you at the door with a schedule change: The 8:45 a.m. standup has been pushed to 9 a.m. and an emergency all-hands added at 11 a.m. She hands you the updated calendar on her phone and you notice she holds eye contact a beat longer than usual, as if checking something. You have been noticing that lately, from her, from several people on the team. A particular quality of attention directed at you whenever AI comes up in any context.

You walk the floor, saying good morning. What you see in their eyes is neither panic nor ease. Your people have read the same articles you woke up to. Their uncertainty starts feeding your guilt. You know they sat in the past through the same meetings where leadership used the words "augmentation" and "evolution" with the careful diction of leaders coached by HR not to say the wrong thing. They are smart enough to hear what is not being said, and they have been waiting for someone to say it with honesty.

You want to reassure them. But you cannot make the promise that reassurance requires. Investors want margin. The math keeps pointing somewhere uncomfortable. You feel that familiar pull between the part of you that is responsible for the balance sheet and the part of you that hired these people, knows their names, knows their family, knows what they are good at and what they are afraid of. You pour a coffee and carry both parts of yourself into the 9 a.m. A thought surfaces, half a sigh, half a gleam, "If only I had known, I would have become a watchmaker."

10:17 a.m.

The first batch of work lands in your inbox and you can feel it before you finish the first paragraph. The report is correct. It is organized. The sentences are grammatically sound. It is also completely hollow. There is no moment in it where a person chose one framing over another because they found it more true, more surprising, more worth saying. There is no voice. There is no risk. The proposal that follows opens with "In today's fast-paced business environment" and you set it down.

You are not angry at the people who produced this. That is the part that takes a moment to work through, because anger would be easier. They felt the pressure to ship faster. The tool was available. You sneer at the irony: You actually shoved it on them. What you are mourning, red pen in hand, is what got exchanged in that transaction: the proof that a person was present, the sentence that went somewhere you did not expect, the evidence of someone who actually cared how the work landed.

You write a note in the margin. You are not sure anyone will understand what you are asking for. You are not even sure the AI will either. Your mind wanders, "Maybe I should phrase the note like a prompt..." You smile at the choice in front of you: passive aggressiveness or patience?

2:34 p.m.

Your stomach has been growling since noon and you have not stopped to snack. You used to have time to eat a real meal. A developer knocks on the open door: The AI-generated module from this morning has a logic error that is cascading through three downstream functions. You pull up a chair. Two hours later, you have fixed it. A mid-level developer writing the original code from scratch would have taken forty minutes.

The productivity argument cited in every vendor deck and executive briefing begins to feel like something someone measured in a laboratory under conditions that do not apply to this office, on this afternoon, with these particular humans and this particular problem.

As hunger and anxiety creep in, you eat another high-protein bar from your desk drawer. You think: "I need to order more. Actually maybe that's a task for a robot or an AI agent. Alexa?"

7:12 p.m.

The house is supposed to be the part of the day that belongs to something else. You remember being a child, spending afternoons with two sticks and an unmanaged imagination, building worlds from nothing, inventing with no screen. You remember dinners that happened without being planned, a table full of friends and family finishing each other's sentences, arguing about the world, someone always about to say something that made everyone laugh. It felt alive.

You're back home, small table tonight: You are four in the same room and, yet three of them are somewhere else entirely. You ask everyone to put their phones away for fifteen minutes while you eat. It feels like you say it the way you would ask for something slightly unreasonable. It takes two reminders to get through the meal. The conversation that follows is good, actually, warm and funny and real, and it lasts about eight minutes before the pull of the devices reasserts itself and everyone drifts.

You stay at the table a little longer than you need to. Looking at the wall. No thoughts. Just peace.

10:48 p.m.

On a comfy couch, TV on, you scroll through what to watch. A thriller about AI surveillance. A series about automated warfare. A documentary about what recommendation algorithms did to a generation of teenagers. You remember seeing films like these decades ago and feeling something close to excitement. They were visions of a future distant enough to be thrilling, the kind of dystopia you could visit and leave. That distance has closed. You are not sure when it happened.

You turn it off. A memory from an old movie, *The Matrix*, surfaces. The way it does sometimes now. Morpheus facing Neo, extending both hands. Red pill, blue pill. When you first saw that image, you knew instantly which to choose. Of course you take the red pill. Of course you want to see how deep the rabbit hole goes. You were the kind of person who stayed curious, who leaned into hard truths, who found discomfort energizing rather than depleting. Cognitively overloaded, you are not sure now.

You go to bed with questions that are not, at their root, about your company's AI strategy. They are about what you are doing with your time, what world you are helping to build, and whether the relentless optimization of everything has quietly removed the things that made it worth optimizing in the first place.

Most of the leaders you know are lying awake with some version of this. None of them talk about it in the morning. You think you should build an association: AI-holics Anonymous. That makes you laugh and cry at the same time. Tomorrow will be the same, rinse and repeat.

This chapter is the conversation they are not having. It starts with a question worth recovering: In times of change, what are we actually trying to preserve?

How We Got There

Culturally conditioned. You never had a chance. A red targeting eye locking onto its mark. A half-human face smiling through the glass with circuitry beneath the skin. The moment the machine turns. We all got conditioned by Hollywood's cautionary tales. They are important, but the real failure mode is slower and organizational. A team so dependent on model outputs that no one retained the habit of questioning them. A product decision that went badly because a system gave a confident answer to launch and nobody thought to push back. Or simply a company that automated through a workforce reduction and discovered two years later that the institutional memory required to course-correct had left the house. Take Disney's pivot away from hand-drawn animation in the early 2000s. They chased the efficiency of CGI, but in the process, they nearly extinguished the institutional soul—the hand-crafted warmth—that made people care about Disney in the first place.

Brain hacked. Notifications, headlines, vendor emails, and competitor announcements are designed to make you feel behind: urgency opens wallets. There is a particular grammar to AI coverage that, once you recognize it, you cannot unsee. A new model releases and headlines declare human-level reasoning achieved. A startup raises a large round and announces it is disrupting an industry it entered eight months ago. A CEO quotes a productivity number on a stage, 40% improvement, costs halved, that no journalist stops to verify and no methodology is ever published. The number travels through social media, earnings calls, and board decks, until it becomes received wisdom.

Consequences. Fear drives the decisions. Many executives are building strategy from a composite picture assembled from press releases, competitor rumors, and conference keynotes. Those sources share one feature: All of them benefit from selling you the AI dream. Disney's early 2000s layoffs unveiled an organization that optimized so aggressively for efficiency that it lost track of what the efficiency was meant to serve. The AI dream is not wrong; the over-promise is.

The Reality

Let's pump the brakes on the marketing rollercoaster for a second: A more nuanced picture emerges. AI tools do accelerate certain tasks. Models can compress research time on well-defined questions and speed up repetitive work. Those gains are real but focused, and often exaggerated outside the context in which they were measured.

Productivity. A 2023 MIT study published in *Science* found that ChatGPT raised productivity among college-educated professionals—time taken decreased by 40% and output quality rose by 18%.[1] That number gets quoted constantly but here is the caveat: The tasks excluded real workplace context. A separate MIT study from 2025 drove the point home more bluntly: Experienced developers using AI wrote code faster but took 19% longer to complete the overall task, and they thought the tool had increased their speed by at least 20%.[2] Confidence outpaced reality.

Cognitive risk. A 2025 study by researchers at Microsoft and Carnegie Mellon University surveyed 319 knowledge workers across real workplace AI use and found that for 40% of tasks, workers applied no critical thinking to AI output.[3] Higher confidence in AI correlated with less thinking. Separately, MIT's Media Lab used EEG to measure brain connectivity during writing tasks and found that LLM assistance produced the weakest neural coupling of any condition tested, weaker than search engines, weaker than working alone.[4]

Security. The Georgetown CSET research confirms AI code generation models frequently output insecure code,[5] a finding consistent with a 2024 peer-reviewed systematic literature review that found approximately 40% of Copilot-generated programs contained vulnerabilities across tested scenarios, with insecure patterns recurring across multiple languages and tools.[6]

Privacy. When Samsung engineers used ChatGPT to debug code and summarize internal meetings, confidential source code entered OpenAI's systems. Cyberhaven's analysis of 1.6 million workers found that 11% of what employees paste into consumer AI tools is confidential.[7] Most of those employees have never read the terms of service that govern where it goes.

Environment. Training GPT-3 emitted roughly 550 tons of CO2.[8] The real cost: Every query consumes electricity, cooling, and water. In 2023, Google reported over six billion gallons of water used in its data centers, driven in part by AI workloads.[9] Deploying AI at scale is both an environmental and operational decision.

Success stories. AI's real value is in its tireless appetite for the boring stuff. Take Walmart's 2023 overhaul of their truck routing. By letting an algorithm obsess over load optimization, they cut $75 million in costs and 72 million pounds of CO_2 emissions.[10] It was a narrow problem met with clean data. Klarna's AI assistant started chewing through two-thirds of their support tickets across twenty-three markets in early 2024. They cut average resolution time from eleven minutes to under two.[11] The focus was on routine queries like order status, payment questions, return policies. Claims about agentic AI are still too fresh to measure long-term impact.

Takeaways. AI works best when the rules are clear and the data matches the real world. Hand it an inhuman volume of similar cases and it moves quickly. Hand it your general infrastructure with poor supervision and you'll end up with a house held together by termites.

Change with Intention

If you don't know where you are going,
any road will get you there.
—Lewis Carroll

Productivity. If a task completes quickly but requires constant oversight and correction, the system shifts the cost downstream. Treat productivity claims like engineers treat benchmarks: Test under real conditions, with real people doing real work. If the gain holds, scale it. If it does not, adjust without embarrassment.

Security and privacy. Anything pasted into an AI tool should be treated as potentially shareable beyond your organization. Establish clear guardrails about what data can be used, where it can be processed, and who remains accountable for the result. Trust in AI systems begins with clarity about boundaries.

Cognitive risk and culture. Keep humans in the loop. A workforce that stops questioning outputs will eventually stop noticing errors. Encourage employees to challenge results, verify assumptions, and explain their reasoning. Double down on critical thinking resilience. AI automation frees time and energy so humans can focus their thinking where it matters most.

Transformation. Keep it simple: Start small, prove value, repeat. Successful AI deployments solve narrow problems first: routing trucks, classifying tickets, reconciling invoices. The work is rarely glamorous, but it is measurable. Choose projects with clear inputs, predictable outputs, and visible return on investment. Deliver small wins, then expand when the evidence supports it.

Leadership. Accountability and trust must not suffer from technological evolution. Successful transitions require adoption with intention: protecting the capabilities that make people valuable while machines handle the rest.

Tomorrow's Will
11:32 p.m.

It is late. The house is quiet in a way it wasn't at dinner. You are still here, still thinking. The guilt you felt looking at your team this morning was the response to a hard situation. You hold onto that. Today will not repeat. Tomorrow will be the change. It will be painful and liberating at the same time. You acknowledge that a small focus team scaling mindfully with AI has an undeniable advantage over a large dragging department.

Perhaps you realized something more important: This transformation forces you to take a step back and look at your company's journey with humility and vision. Maybe you have been postponing looking at your own journey for too long as well. Perhaps it's both. The dinner table with the phones away. The code someone wrote because they cared how it worked. The report that surprised you because a person made a real choice in it. You light up. You had your answer all along. The choice offered to you was about complementarity: AI and human, speed and meaning.

6:15 a.m.

The alarm goes off. The phone is in the same place it was yesterday, and the notifications are already stacking up. But this time, you resist the urge. You sit in the quiet for a moment longer, deciding what parts of the day you're willing to outsource and which parts are non-negotiable.

Today already feels different. You woke up, starting the day on your own terms. You still do not have the full answer. That is fine. The question itself is the thing worth recovering: what to change and what to preserve.

References

1. Shakked Noy and Whitney Zhang, "Experimental Evidence on the Productivity Effects of Generative Artificial Intelligence," *Science* 381, no. 6654 (July 14, 2023): 187–92, https://doi.org/10.1126/science.adh2586.
2. Joel Becker et al., "Measuring the Impact of Early-2025 AI on Experienced Open-Source Developer Productivity," arXiv preprint arXiv:2507.09089, July 2025, https://arxiv.org/abs/2507.09089.
3. Hao-Ping (Hank) Lee et al., "The Impact of Generative AI on Critical Thinking: Self-Reported Reductions in Cognitive Effort and Confidence Effects from a Survey of Knowledge Workers," in *Proceedings of the 2025 CHI Conference on Human Factors in Computing Systems*, article no. 1121 (New York: Association for Computing Machinery, 2025), 1–22, https://doi.org/10.1145/3706598.3713778.
4. Nataliya Kosmyna et al., "Your Brain on ChatGPT: Accumulation of Cognitive Debt When Using an AI Assistant for Essay Writing Task," arXiv preprint arXiv:2506.08872, 2025, https://doi.org/10.48550/arXiv.2506.08872.

5. Jessica Ji, Jenny Jun, Maggie Wu, and Rebecca Gelles, *Cybersecurity Risks of AI-Generated Code* (Washington, DC: Center for Security and Emerging Technology, Georgetown University, November 2024), https://doi.org/10.51593/2023CA010.
6. Claudia Negri-Ribalta et al., "A Systematic Literature Review on the Impact of AI Models on the Security of Code Generation," *Frontiers in Big Data* 7 (2024): 1386720, https://doi.org/10.3389/fdata.2024.1386720.
7. Cameron Coles, "11% of Data Employees Paste into ChatGPT Is Confidential," *Cyberhaven*, February 28, 2023, updated May 19, 2025, https://www.cyberhaven.com/blog/4-2-of-workers-have-pasted-company-data-into-chatgpt.
8. David Patterson et al., "Carbon Emissions and Large Neural Network Training," arXiv preprint arXiv:2104.10350, 2021, https://arxiv.org/abs/2104.10350.
9. Pengfei Li, Jianyi Yang, Mohammad A. Islam, and Shaolei Ren, "Making AI Less 'Thirsty': Uncovering and Addressing the Secret Water Footprint of AI Models," *Communications of the ACM* 68, no. 7 (2025): 54–61, https://doi.org/10.1145/3724499.
10. INFORMS, "Operations Research Enhances Retail Truck Routing and Load Planning: Walmart Awarded the 2023 INFORMS Edelman Award," April 17, 2023, https://www.informs.org/News-Room/INFORMS-Releases/Awards-Releases/Operations-Research-Enhances-Retail-Truck-Routing-and-Load-Planning-Walmart-Awarded-the-2023-INFORMS-Edelman-Award.
11. Klarna, "Klarna AI Assistant Handles Two-Thirds of Customer Service Chats in Its First Month," press release, February 27, 2024, https://www.klarna.com/international/press/klarna-ai-assistant-handles-two-thirds-of-customer-service-chats-in-its-first-month/.

About the Author

William Brendel is a two-time founder and engineering leader with over twenty years of experience in artificial intelligence and product development. He holds a PhD in AI and has authored more than thirty publications and patents, with over 2,100 citations in computer vision and machine learning. He spent a decade at Google, Amazon, and Snap, working at the center of the industry's most significant shifts.

William's philosophy is simple: Build pipelines that handle the drudge work so people can do the inspired work. Today, he applies that approach in HealthTech and renewable energy management.

LinkedIn: https://www.linkedin.com/in/william-brendel/

CHAPTER 6

QUALITY AT MACHINE SPEED: THE FUTURE OF TRUST IN SOFTWARE QUALITY

By Michal Buczko
Software Quality Competence Lead
Wroclaw, Poland

The most dangerous phrase in the language is:
We've always done it this way.

—Grace Hopper

AI has become the accelerator of the entire software lifecycle: research, design, execution, promotion, and maintenance. Work that before took a team weeks to deliver now can be done in an afternoon by one person with an AI model at hand. For example, Amazon recently showcased how six engineers rebuilt a core part of its AI infrastructure in seventy-six days. That work was planned and estimated for a team of forty engineers for a full year in "before AI usage." The code compiles, tests pass, but compiling is not shipping, and passed tests are not the same as quality. AI reduced the cost of producing software code, but it did not lower the cost of making it trustworthy.

The real question has changed and it is no longer if AI can write code, but whether we can trust what AI builds, as fast as it does it. Quality and trust are not a tax on speed. They are the evidence that turns fast-moving AI code into a product you can ship and stand behind. Faster code just moves the slow part downstream; if testing, review, and release cannot keep up, the bottleneck reappears there.

The discipline that closes the gap is orchestration: deliberately deciding who or what does each piece of work, and how to confirm it was done right. That means people, AI models, automated checks, and human review points, coordinated across the whole delivery cycle, not only the coding part. Orchestration, when done well, produces packages and evidence humans can quickly verify, that is, the proof of what was built and is about to ship has earned the trust and quality.

The software delivery lifecycle is now fully AI-touchable and AWS has even formalized this shift as the AI-driven development lifecycle (AI-DLC) framework, in which AI handles much of the execution while humans keep ownership and make final decisions. As figure 1 shows, AI contributes at every phase from research through maintenance.

In this environment each stage that speeds up this way creates a hidden cost: The more output multiplies, the more the risk of quality falls and the effort it takes to keep it. Here is the main focus of quality leaders: Customers do not buy code, but the outcomes they trust to get from it for the price they pay. This is how trust becomes the unit of revenue, so quality cannot report only the test results and bug counts; it has to report the risk, trust, and value the customer can actually experience. The definition of how the trust gets produced and measured is the quality job now.

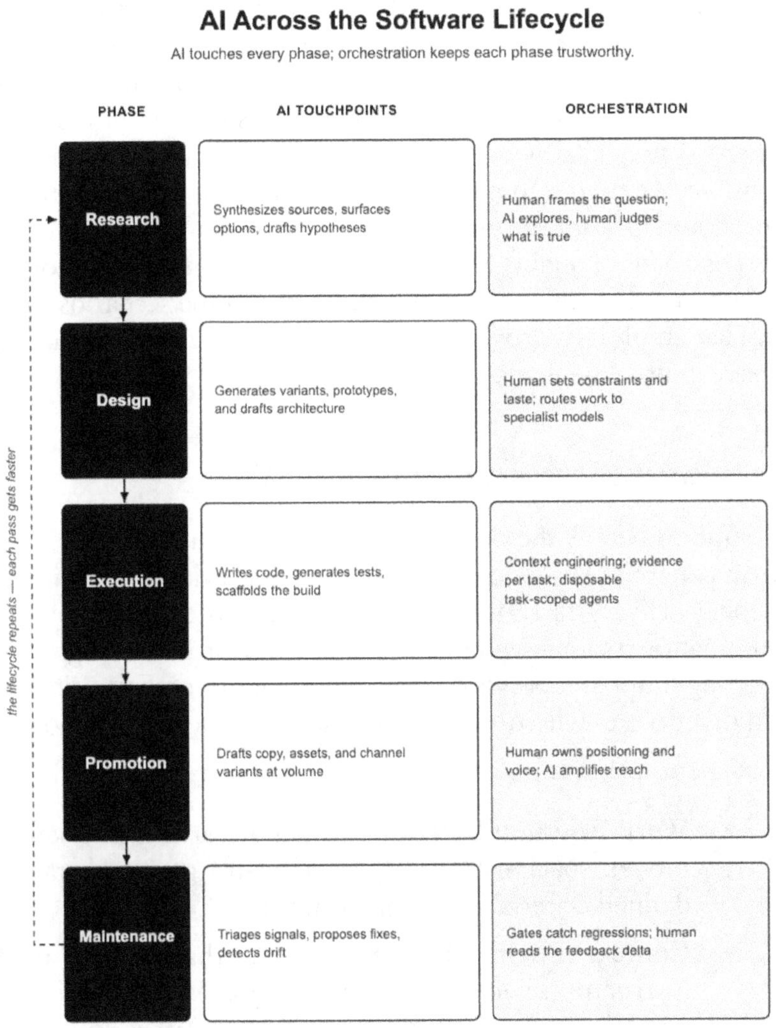

Figure 1. AI touches every phase of the software lifecycle; orchestration is the human discipline that keeps each phase trustworthy.

AI enhancement affects both speed and quality; however, that change is not symmetric. Speed scales very easily; especially as the models grow, quality and trust do not. With the new models released for agentic coding we observe how their capacity is growing and is effectively unlimited, but the human who oversees their operation and delivery—the attention, domain expertise, and judgment of how the pieces integrate together—is not. That is why the bottleneck at the human touch points is so visible. We need to build the ability to verify at speed whether what was produced is fit for the purpose it was designed for. This is a challenge at the crossroads of quality and productivity, pushing quality efforts more to fit-for-purpose verification than exhausting all the scenarios and edge cases. That amplifies a growing need for a framework that builds quality engineering into AI-enhanced delivery.

Orchestration: Deciding Who Does What

Orchestration here is the organization of work routing between people (domain experts), AI models, agents, gates, and evidence, with the goal of keeping delivery fast and verifiable at high speed. It is not focused on automation (acting without a human operator) or augmentation (supporting human operation). In this scenario orchestration is the system that decides who does what and how we know it was done right.

The four main pillars of the orchestration are:

- **Work Routing:** Determine which actor (human, generalist model, specialist model, agent team) gets each task with defined criteria and context limits.
- **Context Engineering:** Everything that model reads is instruction, evidence, or drift. The goal is to feed the first two and starve the third. One important shift is to reorganize around context as a scarce resource, not expertise.
- **Evidence Production:** Define what each piece of work must leave behind, so the next actor can verify it. These can be docs, logs, tests, reviews, and first-class outputs, but not by-products
- **Gate Design:** This is the place where fit for purpose is questioned. Not bureaucracy, but the main contract

between machine speed and human-grade trust. Here KPIs, measurable outcomes, and evidence come together as one package a human can trust.

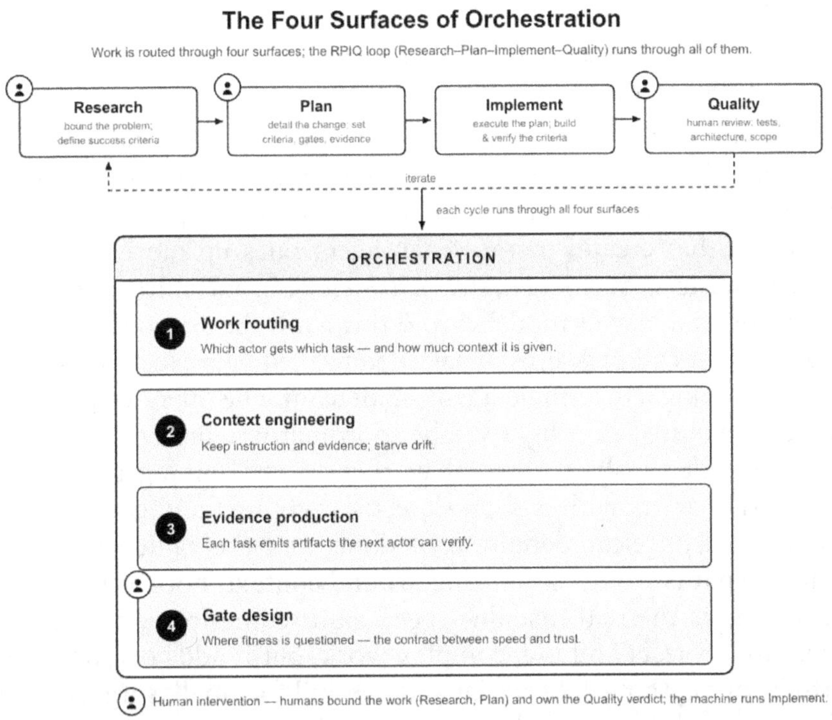

Figure 2. The four surfaces of orchestration, with the research–plan–implement–quality (RPIQ) loop running through them.

The foundation for orchestration is the research, plan, implement, quality (RPIQ) pattern, which has proven to be a very efficient workflow and generalizes well:

- **Research:** Bound the problem and define success criteria.
- **Plan:** Detail the change, write the acceptance criteria, and identify the gates and evidence.

- **Implement:** Execute the plan, implement, build, and validate the acceptance criteria.
- **Quality:** A human reviews to validate that the changes pass all tests, architecture standards, and original scope.

This separates "what the expectation is" from "what will be done" from "what was delivered," three judgments AI models often confuse if the operator gives them opportunity.

Another challenge is the switch from role-based to task-based work. Out of habit, we tend to copy a human work organization onto our AI agents—as we'd hired architects, developers, designers, and testers, but that creates a trap. Each agent piles up more and more information, gets overloaded, strays from its job, and starts interfering with the others. Newer models avoid this, with the approach where a versatile agent can pick up skills needed to the tasks execution on the fly, thus working like a single-agent small team. The operator plans the work and hands it to a lightweight coordinator, which passes small, clearly bounded tasks to several worker-agents running in parallel. Each worker starts with a clean slate, clear context and gets only the skills it needs, a clear definition of done, and the evidence it must produce. That is what "organizing around context, not roles" means, and it's where the real machine-speed gains come from. This shines on mature projects and big, complex work, but it adds overhead. It's worth it for anything substantial and overkill for small, surgical fixes, so keep a simple, direct path for the little changes.

Re-Tooling Quality for Machine Speed

There is no way that the quality practices can keep up if it relies on manual testing alone. Test automation is now an obvious requirement, but even that isn't enough. Quality needs AI's help to keep pace. The efficient path is to take the orchestration we just designed and apply it directly to quality work, so AI can do much of the heavy lifting in quality assurance. To keep up with the speed with AI-driven implementation, quality practices have to start early and carry clear goals and KPIs.

The first quality gate worth prioritizing is static analysis using automated tools that can scan the code without running it. These are well suited for catching what you might call the "AI code signatures." AI-generated code often includes sections that look logical and correct but either never actually run or do nothing useful, along with risks of duplicated helpers or unused declarations. Sometimes the AI doesn't notice the capability already exists and simply builds it again and misses it also during a self-code review. Then these scans, especially the ones that flag unused code and duplication, need to be blocking gates: checks that must pass before work continues, not warnings that could be easily ignored.

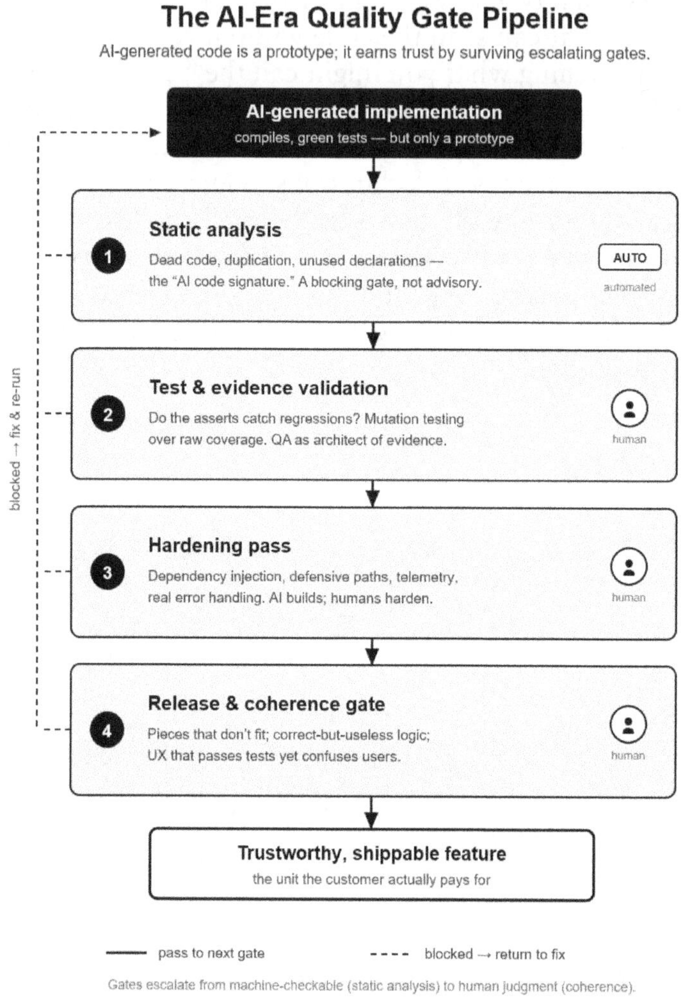

Figure 3. The AI-era quality gate pipeline: AI-generated code is a prototype that earns trust by surviving escalating gates, from machine-checkable static analysis to human coherence review.

Test cases themselves can be designed and generated by AI. In fact, models are very enthusiastic about that work. That means automated testing can be part of the process from day one. But it shifts the quality expert job toward checking the tests themselves. Very often the checks inside them don't actually verify anything meaningful: AI happily produces a wall of passing tests that confirm nothing. This is why the quality expert moves from "quality reviewer" to "architect of evidence," someone who shapes what gets created in parallel to the code and how it gets verified. How much of the code tests cover is a useful starting point, but cannot be the final verdict. It is worth adding mutation testing, which deliberately introduces small bugs to check that the tests actually notice. It is great to have tens of thousands of tests, but you have to keep asking what they really test and whether they would catch something quietly breaking. Otherwise you just pile up tests that cost time to run and maintain.

What AI produces is a great prototype with potential that gets built fast, but it requires hardening, even when it works at first sight. That means production-grade engineering that lets software survive contact with real users: proper error handling, monitoring, sensible fallbacks, and more. AI is great at building the first version, but keeping code healthy as it ages, reworking and hardening it, is where engineers still excel. Designing and overseeing that hardening pass should be the human operator's job.

Before machine-speed code generation, the release gate—the final check before shipping—was mostly about catching the slow trickle of human mistakes. Now it has to catch a different problem: coherence failures, where the parts are each fine in isolation but don't add up. For example:

- Pieces that each look right but don't fit together
- Logic that is mathematically correct but practically useless
- Screens that pass every test yet still confuse customers

The shift can be summarized as a transition from old anchors to new drivers:

Anchor	Driver
Tests as truth	Tests as work products that need checking too
Test coverage as quality	Deep review at sample level
Manual review of everything	Gates as guardrails, evidence as currency
Quality as defender	Quality as architect of evidence
Code is the slow part	Verification is the slow part

Trust Is Produced, Not Felt

Quality and trust have to be merged together, as one part of actual delivery. Together they build a measurable level of confidence that the product will behave as expected and will keep delivering value to the customer. Trust is being built on three pillars: predictability, recovery, and responsiveness. It might dip slightly under the pressure of pace of fast delivery, but only in the limits we expect and measure. What matters is a continuous feedback loop with users, so you can deliver fixes on observed inconsistencies quickly whenever the system behaves unexpectedly. In short, internal quality and maturity measurements have to pivot to practices supporting customer-facing trust contracts. The AI-driven delivery introduces new ways for the trust to potentially fail:

- **Plausibility Drift:** the product looking competent in an area it doesn't actually understand
- **Surface Inconsistency:** different parts of the product doing things differently because they were created in isolation
- **Silent Regressions:** AI fixing one thing and breaks something next to it in a way no test caught

- **Documentation Rot:** generated documentation no longer matching generated code

The quality practice has to find efficient ways to keep these under control when code is being generated at high volume. At that pace, building software starts to be similar to manufacturing production lines, which is exactly why lean quality-control practices get adopted with promising results. Regulated industries—medical devices, automotive, safety critical systems—and the frameworks that govern them demand traceability: a clear record of why each decision was made. For AI-generated work, this traceability is non-negotiable. If you cannot reconstruct why the model built a feature the way it did and justify all the decisions along the way, you cannot ship what it wrote. This is exactly where the orchestration framework fits: Every gate produces evidence, and each gate decision, together with the reasoning behind it, is stored as a record you can audit later.

The Role That Didn't Disappear

From everything above, one thing is clear: Quality practice doesn't disappear; it shifts. The work moves into designing evidence and quality gates, and that requires software engineering skills to optimize context and review evidence, things AI cannot execute for itself. The old-style quality focus optimized on finding defects and executing tests is outdated; the new one utilizes a lean, pragmatic approach to review the right areas in the right moment and build trust in the software.

The hard part is keeping focus and sharp judgement. AI optimizes for answers that look reasonable and logical. Quality has to optimize for what is actually true and check whether "reasonable" is actually true. AI models can mimic expertise in a field but cannot replace knowing that field and customers. That is a quality job now. Same goes with timing: Stepping in at the right moment, with right data and expectation, is itself a required skill. AI excels at drafting, triage, design, translation, refactoring, documentation, and test draft designs; people excel at judging whether the results is a fit for purpose and making the final call. That is why a risk of human bottleneck remains but shifts from project budget and speed to managing limited attention against large workloads. AI does not replace the delivery or quality team members. It accelerates them and changes the team focus that is needed for delivery at AI machine speed.

McKinsey's recent industry-scale research on AI in software delivery showed that as productivity grows in specific kinds of tasks, the human role shifts to orchestration and review. The teams that were most efficient had redesigned their structure and workflow rather than bolting AI onto the old process. Based on the results it was visible how in such teams Agile practices are changing: a movement to small "pizza slice" teams (echo of Amazon's "two pizza teams") where an engineer and AI agent work in short few-day sprints with constant repeating review and feedback loops. They were able to ship production-ready features at a pace not seen before, in a genuinely cross-functional way.

In this new world of AI-enhanced software delivery, quality is no longer about finding bugs or trying to break the software. It is the steering wheel and core skillset of every AI agent operator. The organizations and projects that treat quality and trust as that steering mechanism will outrun the ones that still treat it as a box to tick off at the end. Here is the shape of a new wave of delivery team: human operators working with an AI coordinator agent, surrounded by parallel disposable worker agents. Each one produces evidence and clears a gate at every step, then hands it off to a human engineer at the exact moment when a judgment or decision is required.

So here is what I'd encourage you to try. Take one piece of work where AI is already generating the code, and put a quality discipline around it. Decide what has to be proven before that work can move forward. Define the evidence it must produce, lay out how you'll review that evidence, and choose a short set of unbreakable quality rules that will block a release. Then let AI work at full speed against those rules, and watch what it does. The real lesson isn't in the first attempt. It's in refining those gates, tightening the evidence, and repeating the exercise across the same scope, and later also the rest of your work. That iteration is how quality becomes the thing that steers AI machine-speed delivery, instead of the thing that gets skipped because the code is built so fast.

The teams that thrive won't be the ones with the fastest AI models. It will be the ones who refused to keep delivering the way they always had and made quality their steering wheel in AI-driven development.

About the Author

Michal Buczko is a software quality competence lead with more than twenty years in software testing and quality assurance, currently leading quality across digital and embedded product teams in Carrier HVAC. A member of the European testing community and an international conference speaker, he focuses on how AI is reshaping the economics and discipline of software delivery. He researches how orchestration practices keep AI code built with machine-speed trustworthiness with a design-optimal approach to software quality. Alongside his enterprise work he has spent the last years building a production AI-DLC-driven digital product entirely through AI agent orchestration to validate the approach to quality and trust in AI-built systems.

Email: buczko.michal@gmail.com

Website: https://buczkomichal.ghost.io/

Linkedin: https://www.linkedin.com/in/buczkomichal/

CHAPTER 7

CONNECTING THE MISSING PIECES: AI'S PROMISE FOR MENTAL HEALTHCARE

By Patricia Calazans
AI Researcher, Nonprofit & AI Founder
San Francisco, California

I wish AI could diagnose and treat my condition.

—Daniel Calazans, 2020

As you read this, I want you to imagine sitting beside me, with Daniel's words appearing on my phone screen. Just two weeks before his death in 2020, a friend of his shared a series of messages they had exchanged about artificial intelligence. Daniel has been gone for nearly six years, and reading those messages felt like receiving a letter from the future.

Have you ever wondered what someone close to you would think about the world today? I do, all the time. Daniel worked in robotics, followed emerging technologies closely, and was endlessly fascinated by the possibilities—and consequences—of innovation. So

I find myself asking: What would he think about the extraordinary advances in artificial intelligence taking place now?

Among those messages, one sentence seemed to transcend time: "I wish AI could diagnose and treat my condition."

At first, you might read his message as the optimism of a young technologist. But if you knew what Daniel was living through, as I did, you'd hear something else: the frustration of someone who felt unseen by a system unable to connect the pieces of his story.

Daniel understood what many patients discover too late. Mental health symptoms are often treated in isolation, disconnected from the biological, neurological, and social realities that may be driving them. What he wanted was not a machine to replace human care.

He wanted a system capable of understanding the whole person. That distinction matters—to Daniel, to me, and maybe to you, too.

Today, artificial intelligence is rapidly entering mental healthcare, not because the field is fully prepared, but because the need is overwhelming. Long wait times, clinician shortages, rising costs, and persistent stigma have driven millions toward AI-powered tools that offer something traditional systems often cannot: immediate access.

The promise is enormous. So are the risks. Let's be honest with each other: Hope and caution have to walk side by side. The future of AI in mental health will not be determined by how quickly technology advances, but by whether we use it to strengthen human care or simply to scale the limitations of existing systems.

A System That Never Saw the Whole Person (And Why That Matters to All of Us)

In February 2020, Daniel survived a fentanyl poisoning.

The pill he took contained fentanyl without his knowledge. He collapsed on his way to work and survived only because he fell within sight of paramedics, who administered naloxone and intubated him on the sidewalk.

The medical emergency lasted hours. Its consequences lasted months.

Research now shows that non-fatal opioid poisonings can result in hypoxic brain injuries, which affect memory, emotional regulation, anxiety, and depression. Daniel also suffered significant kidney damage that required hemodialysis, a condition independently associated with elevated rates of depression and suicide risk.

Adding to that burden, he survived a serious assault during a rehabilitation stay, an event that never followed him through the healthcare system because it was omitted from his discharge records.

Each specialist saw a piece of the puzzle. No one saw the whole picture. What followed was largely interpreted as a psychiatric illness. Yet many of Daniel's symptoms were consistent with the combined effects of neurological injury, renal disease, trauma, medication exposure, and psychological distress.

The question that was never systematically asked was the one that mattered most: What is causing these symptoms?

In the weeks following Daniel's death, I did what any desperate parent might do: I searched for answers the only way I knew how: as a journalist. I spoke with physicians, mental health professionals, researchers, and parents who had lost children of their own. Again and again, I encountered the same pattern: fragmented care, disconnected records, and a system that often treated symptoms without fully investigating underlying causes.

That work led me to establish the Daniel Calazans Foundation, dedicated to raising awareness of the relationship between mental health, substance use, trauma, and medical conditions that can mimic psychiatric disorders.

Only later did I realize that the type of system Daniel needed—a system capable of connecting information across providers, specialties, and diagnoses—is precisely where artificial intelligence may offer its greatest contribution. Ironically, those are not the AI applications attracting the most attention or investment today.

A Crisis Larger Than Technology

The mental health crisis is impossible to ignore. According to the 2024 Lancet Psychiatry Commission, rates of mental illness among young people have increased dramatically over the past two decades. Mental health conditions now account for nearly half of the disease burden experienced by youth globally, yet receive only a fraction of health funding.

In the United States alone, more than 49,000 people died by suicide in 2023. Millions more reported serious suicidal thoughts, and millions received no care at all. At the same time, hundreds of thousands of families have lost loved ones to opioid overdoses. These numbers represent more than a healthcare challenge. They represent a failure of systems. And fragmented systems are precisely the kind of problem artificial intelligence was designed to address.

There is another reason this conversation matters. For the first time in history, millions of people are turning to artificial intelligence not only for information, but for emotional support. Young people increasingly use AI to ask questions they are unwilling to ask parents, teachers, clinicians, or friends. They seek advice about anxiety, depression, relationships, loneliness, purpose, and identity. This trend should not be dismissed. It reveals both the accessibility of AI and the gaps in our existing systems of care.

Yet it also raises an uncomfortable question: If people are turning to machines for support, what does that tell us about the systems designed to support them? The growing popularity of AI in mental health is not merely a technology story. It is evidence of unmet human needs. Understanding those needs may be just as important as improving the technology itself.

Yet much of the public conversation around AI and mental health has focused on chatbots. Can they provide therapy? Can they reduce costs? Can they replace clinicians? These questions matter, but let's not get distracted. There's a far more transformative opportunity here, one I want to explore with you. The greatest potential of AI may not lie in simulating empathy. It may lie in helping clinicians identify what humans routinely miss.

Artificial intelligence can analyze enormous volumes of medical information, identify hidden patterns, monitor medication interactions, integrate fragmented records, and flag risks that might otherwise go unnoticed. In other words: AI might help most not by pretending to be a therapist, but by helping real clinicians give better care.

Five Ways AI Could Have Changed Daniel's Outcome

Daniel's case was not clinically mysterious. The information existed. What was missing was a mechanism capable of integrating that information quickly enough for action. Today, several categories of AI tools already exist that could have altered critical points in his care journey.

First, AI-powered clinical decision support systems can identify patients at risk for neurological complications following hypoxic events and automatically recommend follow-up evaluations.

Second, AI can support multidisciplinary care coordination by identifying patients whose conditions require intervention across multiple specialties.

Third, validated predictive models can identify elevated suicide risk among patients with chronic illnesses such as kidney disease, allowing for earlier monitoring and intervention.

Fourth, AI-driven pharmacovigilance systems can evaluate medication interactions, contraindications, and cumulative risks in ways that surpass the capacity of any individual clinician.

Fifth, interoperability systems can help ensure that critical information follows the patient rather than disappearing among providers, facilities, and medical records.

None of these capabilities requires science fiction. They already exist in various forms today. What's missing isn't the technology. It's something more basic and harder to build. It is the priority, investment, and regulatory commitment necessary to deploy it at scale.

Making the Invisible Visible

During treatment, Daniel once told me he wished he had cancer. It wasn't because he wanted to be sicker. Maybe, in a way, you can relate to that wish for certainty, even if you've never said it out loud. It was because cancer appeared on a scan. It produces evidence that clinicians can see and measure, something to point to, something to rally around when organizing treatment.

Mental illness often does not. Psychiatric diagnosis still relies heavily on clinical evaluation, self-reporting, and subjective interpretation. Symptoms frequently overlap. Underlying causes are often difficult to identify. This is where AI may prove transformative.

Researchers are increasingly using artificial intelligence to analyze neuroimaging data, identify patterns associated with depression and addiction, and detect risks earlier than traditional approaches allow. The goal is not to replace clinicians. The goal is to help them see what has historically remained hidden. The future is not AI pretending to be a therapist. The future is AI helping clinicians connect what fragmented systems have failed to recognize.

The Horizon Ahead

Much discussion about artificial general intelligence focuses on automation and economic disruption. For mental health, the more important question may be integration.

Today's AI systems excel at narrow tasks. One model may identify medication risks. Another may detect neurological abnormalities. Another may predict suicide risk. What they cannot yet do is reason across all these domains simultaneously.

If more advanced forms of AI eventually emerge, their greatest contribution to healthcare may not be replacing physicians but helping them synthesize complex information across specialties. The defining failure in Daniel's case was not a lack of expertise. It was the absence of synthesis.

His brain injury, kidney disease, trauma history, medication regimen, and psychiatric symptoms existed in separate records interpreted through separate lenses. The promise of AI is the ability to connect those domains into a unified understanding of risk. That promise, however, comes with responsibility.

AI must be developed with scientific rigor, transparency, accountability, and human oversight. Systems that inherit the biases and fragmentation of existing healthcare structures will not solve the problem. They will simply automate it.

Throughout my career in communications, I have learned that every major technological shift eventually becomes a trust issue. The internet changed how we access information. Social media changed how we communicate. Artificial intelligence is beginning to change how we think, decide, learn, and seek support.

The question isn't whether AI will shape the future of mental health. It already is. But will it help us see each other more clearly or just speed up what's already broken? It already does.

The real question is whether society can build enough trust, oversight, transparency, and accountability to ensure these systems serve people rather than simply engage them. Mental health is among the most sensitive areas in which AI will operate, as it touches on identity, vulnerability, suffering, and hope.

When the stakes involve human lives, innovation alone is not enough. Trust must become part of the design.

The Answer to His Question

Artificial intelligence cannot replace compassion, clinical judgment, responsibility, or human connection. It cannot heal grief. It cannot love. But it can help reveal patterns that have remained hidden for far too long. It can help connect information that never should have been separated. It can help clinicians recognize risks before they become tragedies. Most importantly, it can help us move toward a future where fewer families are left asking the questions mine still carries.

Daniel did not need a chatbot. He needed a system capable of seeing him completely. Millions of people need the same thing today.

Whether artificial intelligence becomes another layer of fragmentation or a force that helps us finally understand the whole person depends on the choices we make now.

The answer is ours to write.

Keep questioning.

About the Author

Patricia Calazans is a communications strategist, AI researcher, and author. She completed advanced studies in artificial intelligence at the University of California, Berkeley and Stanford University. Her work explores the intersection of AI, mental health, healthcare, and public trust.

LinkedIn: https://www.linkedin.com/in/patriciacalazansashton

CHAPTER 8

THE FOUR PILLARS OF THE AI-AUGMENTED SELLER: A CONSTRUCTIVIST GUIDE TO NEXT GENERATION SELLING

By Jeffrey C. Granvold
Senior Technology Strategist, Author, AI Leader
Dallas, Texas

The illiterate of the 21st century will not be those who cannot read and write, but those who cannot learn, unlearn, and relearn.

—Alvin Toffler

Who remembers their first encounter with the concept of "the Cloud"? It was a nebulous, almost mystical promise that our data could live everywhere and nowhere at once. That transition wasn't just a technical upgrade; it was a psychological leap of faith. It required us to trust that our most valuable assets were safe in a space we couldn't touch, shifting the paradigm from ownership to agility.

We see this same leap of faith today with artificial intelligence. We are inundated with narratives about how AI will reshape society, yet the reality remains: AI cannot replace the human mind's ability to construct deep, meaningful relationships. This is where the principles of constructivism become essential.

Constructivism is a psychological and educational theory which posits that individuals do not passively receive information, but instead actively construct their own knowledge and reality based on their unique experiences and social interactions. As noted by the influential theorist Jean Piaget: "Knowledge is not a copy of reality. To know an object, to know an event, is not simply to look at it and make a mental copy or image of it. To know an object is to act on it."

My professional foundation was built on this very philosophy. Before leading sales strategies at Verizon Business or consulting on IT optimization with my clients, my world was shaped by social work and professional counseling. The transition from clinical work to IT sales was less of a climb than it might seem; both require a deep, empathetic understanding of how people use the tools at their disposal to make decisions.

This curiosity about the human "why" is part of my DNA, inherited from my father who was a professor and visionary in the field of constructivist counseling. He instilled in me a lifelong commitment to understanding the frameworks of the mind. Through this lens, I view AI not as a replacement for the seller, but as a cognitive scaffold. It is a support structure that allows us to reclaim the human premium in our work.

In the hands of a next-generation salesperson, AI serves to uncover the hidden context of a buyer's world, transforming raw data into a shared discovery space. Here, the salesperson doesn't just deliver information; they facilitate a "reflective dialogue" that helps the client reorganize their own business reality. By integrating AI-driven insights with empathetic inquiry, the seller bridges the gap between current challenges and a more efficient future. The result is not merely a transaction, but a meaningful co-creation built on mutual trust.

To thrive in this era, professionals must rebuild their frame of mind around how AI enhances the four primary pillars of sales excellence:

1. **Prospecting:** Moving from cold outreach to informed entry
2. **The Client Relationship:** Shifting from vendor to trusted advisor
3. **Ecosystem Excellence:** Creating alignment through shared data insights
4. **Time Management:** Automating the administrative to prioritize the interpersonal

Prospecting

Prospecting is one of the skills that got you hired, yet in the age of AI, the nature of that skill is undergoing a radical transformation. Chris Luxford, a twenty-five-year management consultant and author of leadership strategies, suggests that we are currently facing an extinction event for those who rely on the traditional, high-volume numbers game of generic outreach. Applying this to a constructivist framework with AI, you must reimagine the proactive creation of opportunity based not just on your outcomes, but on how those outcomes changed your point of view relative to the business problem.

The Scenario: Reimagining a Stagnant Retail Account

Consider a senior IT sales executive targeting a large retail chain struggling with dead stock (items that don't sell) in some regions while facing stock-outs (empty shelves) in others.

- **The Traditional Approach (The Product-Pusher):** The seller focuses on technical specs like barcode accuracy. The value proposition is a technical upgrade, but the business problem—misallocated capital—remains unaddressed.

- **The Next-Gen Approach (The Ideas Factory):** Using AI to synthesize market data, the seller discovers the retailer's inventory ignores local micro-trends.

The Strategy:

- **Constructing the Outcome:** Before the meeting, the seller uses AI to model how regional economic shifts affect demand, identifying that southern regions need a different inventory cadence than the north.
- **The Shared Discovery Space:** The seller presents a vision of "living inventory," predicting demand at the zip-code level.
- **Creating the Visionary:** The retail executive shifts from a push model to a pull model. The seller provides the cognitive scaffold for the client to become an innovator in precision retail.

The Client Relationship

The second pillar shifts the focus from transaction to partnership through reflective dialogue. With AI managing the administrative memory—tracking the nuances of past conversations and shifting priorities—the seller is freed to engage in deep empathy and complex problem-solving.

The Scenario: Navigating a Shifting Corporate Strategy

Imagine you are a sales professional working with a long-term client at a large logistics firm who has an upcoming renewal for their CRM licenses. In past conversations they shared their interest in sustainable packaging as part of a shift in corporate strategy.

- **The Legacy CRM Experience (The Filing Cabinet):** In the old model, your call is focused solely on the renewal. You have forgotten that six months ago, they mentioned an interest in sustainable packaging. The call is transactional and you are categorized as a typical vendor focused solely on the next sale.
- **The Next-Generation Experience (The Reflective Dialogue):** AI surfaces a nuanced insight: The client's priorities have shifted from speed to sustainability.

The Strategy:

- **Deep Empathy:** You lead with the human "why," asking how new carbon regulations are affecting their team's ability to execute.
- **The Co-Creation:** You use AI to pull up a future state model of how your technology tracks carbon offsets.
- **The Result:** You aren't just another vendor; you are a partner co-constructing together a solution for their new reality while also becoming "part of the fabric" of your customer's operations organization.

Ecosystem Excellence

Our professional reality is defined by the networks we build. My own perspective was shaped early in my career by a pioneer in ecosystem management who taught me that when partners work together, they change the very nature of how we serve a client. This "ecosystem-first" mindset is now the foundation for the next generation of sales. Today, being a master of the partner network isn't enough; you must also master the agentic tools that drive modern engagement. As Marc Benioff, CEO of Salesforce shared in January 2026, humans must remain at the center of our tech-driven world because only people can "see around corners." In this new agentic enterprise, the salesperson's role has reached its highest form: moving beyond the transaction to become a master orchestrator of value.[1]

The Scenario: The Frictionless Retail Transformation

A retailer wants a "just-in-time" inventory model, but their infrastructure is fragmented between legacy systems and localized data silos. This is preventing real-time, end-to-end visibility to synchronize supply with immediate demand.

The Orchestrator's Strategic Framework

- **Constructing the Force Multipliers:** Rather than selling a product, the orchestrator assembles a high-impact group of partners—a connectivity specialist (5G/IoT), a strategic logistics consultant, and a data integration expert—to heal the retailer's fragmented legacy patchwork.
- **AI-Driven Diagnostics:** Agentic AI scans the blind spots of the supply chain, revealing a critical leak: The retailer is losing 4% of perishable inventory because disconnected trucks fail to communicate environmental data in real time.
- **The Reflective Dialogue:** You facilitate a shift in perspective from "buying tech" to "Holistic Product Integrity," leading a session where stakeholders see how synchronized sensors and predictive routing transform a fragmented loss into a unified gain.
- **Co-Creating the Agentic Enterprise:** The outcome is a unified ecosystem proposal where agentic AI acts as the connective brain, autonomously orchestrating human partners and machine intelligence to achieve a true, waste-free just-in-time model.

Time Management

"Every time you can automate routine but disciplined work, you're moving the needle toward having more capacity to do creative problem-solving work. That's where you get the power and real promise of what AI is—people doing the best work of their life."

—Barry O'Reilly

In December of 2025, I wrote an article on LinkedIn comparing Jim Croce's song, Time in a Bottle comparing it to how AI reclaims humanity. Before AI, the traditional salesperson was often viewed as the "volume engine" where success was measured by the sheer endurance of the hustle. A salesperson's time management was a battle against the clock to squeeze out "more": more cold calls, more emails, and more manual data entry. In this legacy model, the seller's value was inextricably tied to their capacity for repetitive labor. However, as we transition into the era of AI-augmented selling, the role of the salesperson evolves from a processor of tasks to a strategist of intent. This shift represents a profound psychological pivot: Time management is no longer about the rigid management of minutes, but the deliberate stewardship of cognitive energy. The future seller is not defined by the quantity of their activity, but by the quality of their presence, leveraging technology to offload the mechanical so they can master the meaningful.

The Science of Cognitive Offloading

The modern seller faces a constant state of administrative friction. Every CRM update, scheduling conflict, and manual follow-up consumes a portion of our finite daily willpower. Through cognitive offloading, we are not just delegating tasks; we are preserving our "working memory" for high-stakes problem-solving. By offloading the "work about the work" to AI, we mitigate decision fatigue, ensuring that our highest quality of thought is reserved for the client's most complex challenges.

The Case Study: Reclaiming the "Golden Hour"

The true value of AI is best seen through the lens of the golden hour, that peak window of cognitive clarity, usually in the early morning, where our ability to synthesize information and recognize patterns is at its zenith.

1. The Traditional Workflow (Cognitive Overload):

In the legacy model, a seller spends their peak morning hours wrestling with administrative noise. They are reactive, responding to emails and manually updating deal stages. By the time they meet a client for a white boarding strategy session in the afternoon, their mental "battery" is depleted. They are physically present but cognitively exhausted, leading to surface-level contributions and missed nuances.

2. The Next-Gen Workflow (Cognitive Offloading):

- **The Autonomous Offload:** While the seller rests, AI agents act as a digital "chief of staff." They synthesize transcripts from the previous day's meetings, highlight action items, and prepopulate CRM fields.
- **The High-Value Reclamation:** Instead of starting the day in a reactive state, the seller enters their morning in a state of deep learning. With administrative friction removed, they spend their golden hour researching the client's long-term sustainability goals or dissecting an annual report to find hidden points of leverage.
- **The Strategic Performance:** The seller arrives at the board meeting not as a vendor, but as a consultant. Because their mental bandwidth is fresh, they can pivot in real

time, framing technology not as a product, but as the foundational layer for the client's future legacy.

From Data Prisoner to Value Architect

Ultimately, this evolution in time management marks the transition from being a prisoner of data to being an architect of value. We often mistake being busy for being productive, but AI forces us to confront the difference. When we offload the noise, we don't just gain more hours; we gain a higher quality of presence. When we protect our cognitive energy, we don't just work smarter. We show up more authentically. We transform the sales process from a relentless chase into a series of meaningful, human-centric encounters. In this new era, the greatest ROI of AI isn't the time it saves, but the humanity it restores.

References

1. Alvin Toffler, "The Illiterate of the 21st Century..." Forbes, October 04, 2022, https://www.forbes.com/sites/eliamdur/2022/10/04/the-illiterate-of-the-21st-century/
2. Piaget, J. (1964), "Part I: Cognitive Development in Children: Piaget Development and Learning." *Journal of Research in Science Teaching*, vol. 2, no. 3, 1964, pp. 176–186. *Wiley Online Library*, https://doi.org/10.1002/tea.3660020306.
3. Chris Luxford, "The Burden of Growth Without Control: A Call to Arms for Sales Leaders" December 8, 2025, https://medium.com/@cluxford/the-burden-of-growth-without-control-a-call-to-arms-for-sales-leaders-ce9080cb4330
4. Marc Benioff, "The Truth About AI" *TIME*, January 15, 2026, https://time.com/collections/davos-2026/7339209/ai-revolution-agentic-enterprise/

5. Jeffrey Granvold, “Time in a Bottle: How to Solve Time Management in an AI World”. December 27, 2025, https://www.linkedin.com/pulse/time-bottle-how-solve-management-ai-world-jeffrey-granvold-ygidc/

About the Author

Jeffrey Granvold is a senior technology strategist, author, and presenter with more than 25 years of experience at the intersection of sales leadership and technical innovation. During his 14-year tenure in leadership at Verizon Business, Jeffrey became a leading voice in the disruption of customer experience technologies, building sales organizations focused on outcomes-based selling and early adoption of AI.

Jeffrey is also a combat veteran of the United States Navy, having served as an operations specialist. He was awarded the Southwest Asia Medal (with two Bronze Stars), Kuwait Liberation Medal, Navy Achievement Medal, and the Combat Action Ribbon. He holds an MBA and a Bachelor's in Social Work from Loyola University Chicago.

LinkedIn: https://www.linkedin.com/in/jeffreygranvold/

CHAPTER 9

CLOSING THE AI ADOPTION GAP: THE HUMAN FIX NOBODY'S SELLING YOU

By Scott Hall
AI Transformation Leader, 5-time Author
New York, New York

The future isn't AI versus humans. It's AI plus humans
—Shep Hyken, CX expert

Your organization purchased cutting-edge AI software. Consultants arrived with presentations demonstrating how tasks that normally consumed weeks could happen in minutes. The executive team got energized. Budgets got approved. Company-wide emails announced the future had arrived.

Then nothing happened. Your top performers began updating their LinkedIn profiles. Team members cracked nervous jokes about being replaced by algorithms. Those expensive licenses are sitting unused while everyone claims they're too swamped to learn something new.

Your AI implementation isn't failing because you chose the wrong platform. In fact, the majority of AI initiatives collapse before hitting the ninety-day mark. Why? The reason isn't technical, it's human. People will not transform their work habits just because you installed new software.

The AI adoption gap is the chasm between having AI technology and actually getting people to use it effectively. Ninety-five percent of AI projects fail to generate any meaningful value, according to MIT Media Lab research. Oftentimes, these initiatives turn into expensive obstacles, burning through six or seven figures before quietly getting shelved.

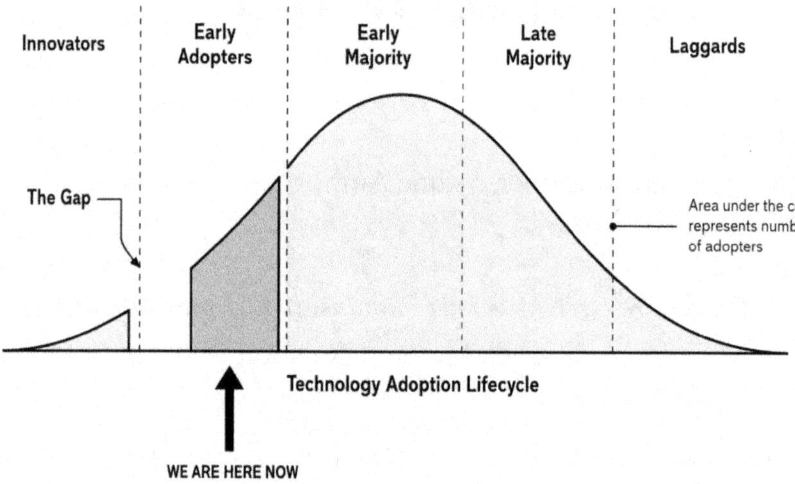

The Unstable Foundation Most Companies Are Building On

Organizations approaching AI concentrate on two elements: strategic direction (identifying which processes to enhance) and technical execution (deploying systems and ensuring mechanics function). Since AI became mainstream in late 2022, businesses have poured resources into these areas. Then reality arrived.

The foundation collapsed because the third pillar didn't exist: the human dimension. This is where organizational behavior, executive alignment, and transformation management become non-negotiable. Most executives don't possess frameworks to shepherd organizations through this fundamental shift. They skip the human pillar entirely and can't figure out why investments aren't paying off.

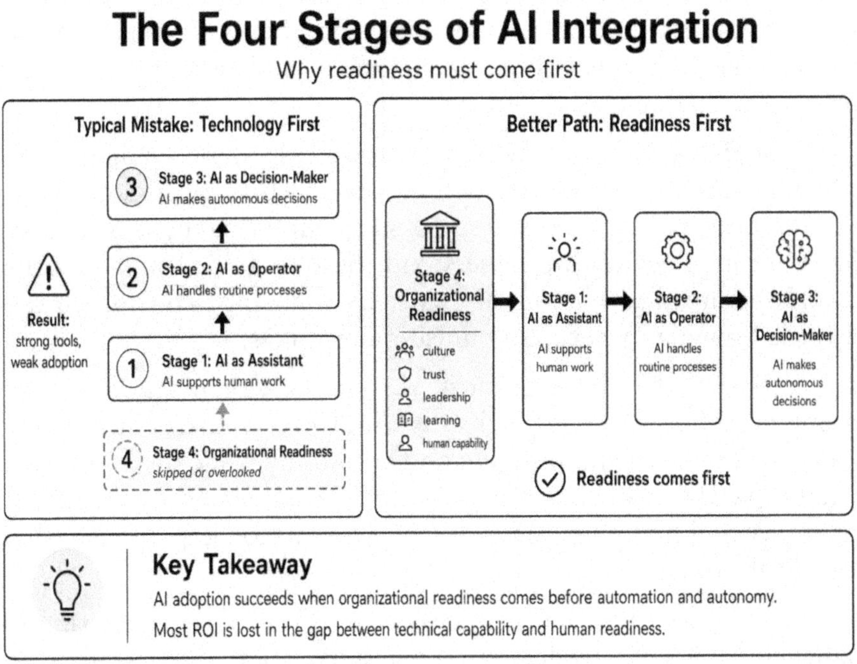

Most Organizations Are Moving Through AI Integration Completely Backwards

Walk into any consulting firm and they'll present you with essentially the same roadmap. It appears logical, makes intuitive sense, and completely misrepresents how AI actually gets adopted in functioning organizations.

Stage 1 is AI as assistant, where people handle the work while AI provides suggestions and enhancements.

Stage 2 is AI as operator, where AI executes routine processes independently while humans manage oversight and exceptions.

Stage 3 is AI as decision-maker, where AI functions autonomously, executing decisions and actions without continuous human intervention.

Sounds reasonable, right? Start with simple support, build organizational confidence, then scale complexity. Except there's a fourth stage that gets completely ignored. And without it, the first three stages are guaranteed to fail.

Stage 4 is organizational readiness. Your company has cultivated the culture, established the trust networks, and developed the human capabilities required to successfully navigate stages one through three. Leadership understands how to guide transformation. Teams operate in psychologically safe environments. The organization rewards learning rather than punishing mistakes.

Here's what destroys most AI rollouts: Organizations attempt to implement stages 1 through 3 without ever constructing stage 4. They leap directly to automation and autonomy while their workforce remains paralyzed by uncertainty and fear. That gap between technical capability and human readiness is the AI adoption gap and it's where your ROI disappears.

AI adoption success depends 80% on transformation management and leadership capability, not on strategy or technology. While strategic planning defines objectives and technical implementation defines mechanics, transformation management determines whether people actually participate. When employees don't engage with AI systems, no investment in strategy or technology delivers value.

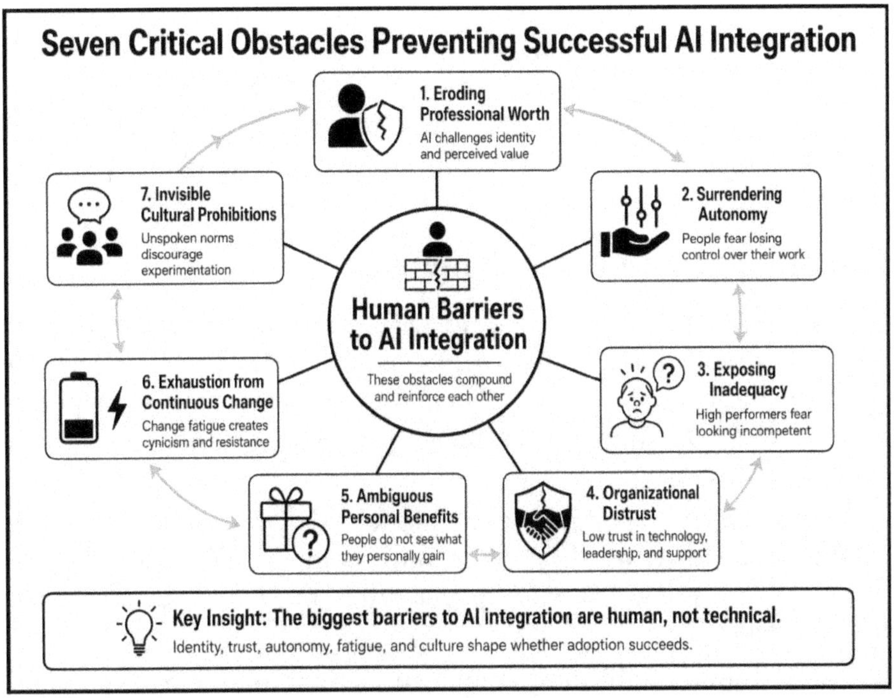

What's Actually Blocking Successful AI Integration

Before you can build organizational readiness, you need to understand what's actually blocking it. Most companies focus on technical obstacles, but the real barriers are human, and they're remarkably consistent across industries.

Obstacle 1: Eroding Professional Worth

Employees have invested years constructing specialized knowledge that anchors their professional identity. When AI executes their signature work instantaneously, it fundamentally challenges who they believe themselves to be.

Consider the marketing strategist who built their reputation on campaign insights requiring deep market understanding. When AI generates comparable strategies in moments, their unique value proposition evaporates. They're not merely concerned about job security but about professional relevance itself. Without helping people reconstruct their value around strategic judgment and contextual wisdom rather than technical execution, resistance becomes inevitable.

Obstacle 2: Surrendering Autonomy

Professionals associate autonomy with competence and status. AI injects unpredictability into workflows that were previously under complete individual control. When someone personally constructs a financial model, they oversee every assumption, every calculation, every conclusion.

When AI produces that model, they're validating work generated by systems they don't fully comprehend. The question becomes paralyzing: "What happens when the AI makes an error I fail to detect?"

Without establishing frameworks that preserve meaningful autonomy within AI-enhanced processes, people will either reject the technology entirely or misuse it to maintain their illusion of control.

Obstacle 3: Exposing Inadequacy

High-performers especially dread becoming the person who struggles with new systems. AI adoption creates temporary incompetence for previously capable professionals. They're unfamiliar with the vocabulary. They don't recognize effective interaction patterns. They cannot distinguish reliable outputs from sophisticated fabrications. In competitive environments where confusion signals weakness, this anxiety intensifies.

People will pretend comprehension rather than seek clarification, guaranteeing poor implementation and eventual abandonment of the tools.

Obstacle 4: Organizational Distrust

Distrust manifests across multiple dimensions, and weakness at any level generates resistance.

Technological Trust: "Will these systems perform reliably or produce errors that undermine my professional credibility?"

Leadership Trust: "Is management deploying this to enhance my capabilities or to document my obsolescence?"

Institutional Trust: "When I inevitably make mistakes during this learning process, will the organization support my development or use those failures against me?"

When trust breaks down at any dimension, people engage defensively rather than openly, never developing genuine capability.

Obstacle 5: Ambiguous Personal Benefits

People require a clear understanding of what they personally gain that justifies the disruption of changing established workflows.

Most AI implementations emphasize organizational advantages like operational efficiency, reduced costs, and market positioning. What remains unclear is the individual value proposition: "Does this technology liberate me from tedious work I despise so I can focus on intellectually stimulating challenges? Does it elevate my organizational value by enabling higher-level strategic contribution? Does it reduce professional stress by handling routine decisions?"

Without articulating compelling personal benefits, AI adoption appears as additional effort benefiting everyone except the people doing the work.

Obstacle 6: Exhaustion from Continuous Change

This represents neither the first nor likely the second major organizational transformation your company has launched recently. Employees are depleted from previous initiatives announced enthusiastically, requiring substantial adaptation, then quietly abandoned when results are disappointing.

Or worse, initiatives that succeed organizationally increase individual workload. This exhaustion breeds cynicism. People hear "AI transformation" and mentally translate it to "another temporary disruption we'll outlast."

Overcoming this requires acknowledging the fatigue explicitly, demonstrating why this transformation differs fundamentally, and proving through tangible early victories that this change genuinely improves individual work experiences.

Obstacle 7: Invisible Cultural Prohibitions

Even when individuals recognize AI's potential value, they frequently resist adoption because of unspoken organizational norms defining acceptable behavior.

In certain cultures, utilizing AI assistance carries subtle implications of intellectual inadequacy or professional shortcuts. In others, early adopters become organizational guinea pigs who absorb blame when implementations falter.

People conform to peer expectations even when those expectations damage organizational progress. Transforming these norms demands visible leadership adoption, credible peer advocates demonstrating success, and explicit organizational permission to experiment without professional consequences.

These seven barriers don't exist in isolation. They compound and reinforce each other. Someone experiencing identity threat is more likely to distrust leadership's intentions. Someone with competence anxiety is more susceptible to peer pressure against adoption.

These seven obstacles operate interdependently, compounding and amplifying each other. Someone experiencing eroded professional worth becomes more susceptible to organizational distrust. Someone anxious about exposing inadequacy proves more vulnerable to invisible cultural prohibitions against adoption.

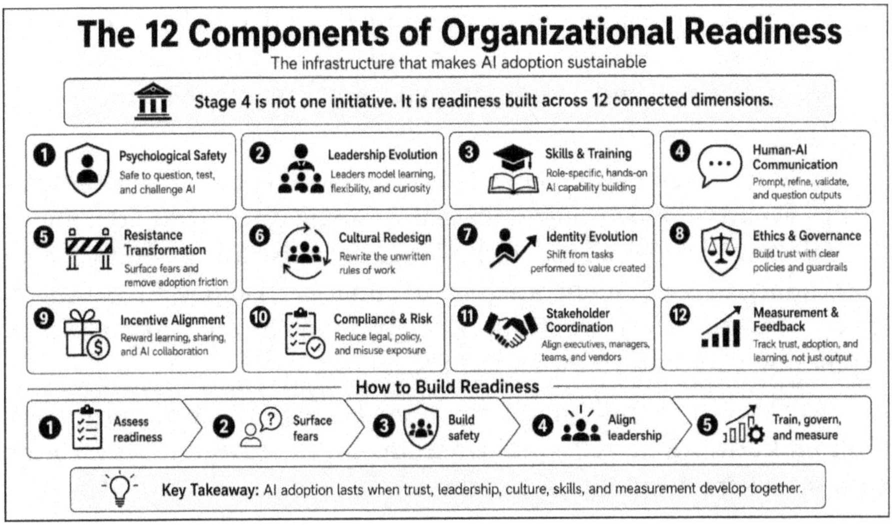

What It Really Takes to Make an Organization Ready for AI: The Building Blocks

Building stage 4 isn't a single initiative. It's comprehensive development across 12 interconnected dimensions. Consider these the infrastructure making AI adoption sustainable rather than a temporary experiment.

Dimension 1: Human Preparedness and Psychological Security

Everything originates here. To enable cultural transformation, you must address fear and perceived loss of control. You must address identity threats and establish emotional security for experimentation.

Without psychological security, people won't reveal confusion, won't experiment with unfamiliar approaches, and won't challenge clearly incorrect AI outputs.

Dimension 2: Leadership and Mindset Evolution

Leaders must transition from traditional models where authority and expertise were concentrated in leadership to new models where leaders orchestrate and coordinate expertise rather than being the sole expert.

Simultaneously, this involves transforming cultural paradigms from scarcity and control toward abundance and flexibility. Leaders must model adaptive thinking, demonstrating comfort with uncertainty and willingness to learn publicly. Training in paradigm transformation helps establish curiosity foundations within culture, promoting exploration rather than resistance.

This creates discomfort for many leaders whose authority was built on knowing more than their team, but it's essential for creating cultures where AI adoption can thrive.

Dimension 3: Practical Skills Development and Training

People need concrete capabilities to work effectively with AI. This includes understanding how to craft effective prompts, recognize hallucinations and bias, validate outputs, and integrate AI into their specific workflows.

Training must be role-specific and hands-on, not generic overviews. A marketing team needs different AI skills than a finance team. A legal department requires different competencies than customer service.

Without practical, contextualized skill development, people remain perpetually incompetent and frustrated, never moving beyond surface-level experimentation.

Dimension 4: Human-AI Communication Frameworks

Teams need developing emotionally intelligent language for interacting with AI systems. This transcends simple prompting and involves creating communication frameworks, understanding which inputs to trust, recognizing biases, and effectively translating AI responses.

People must develop skills to question AI outputs, identify when results are unreliable, and iteratively refine their requests to get better outcomes. This is about developing genuine conversational competency with intelligent systems.

Dimension 5: Resistance Analysis and Transformation

Teaching teams to decode underlying fears, eliminate friction, and create pathways beyond resistance is essential. Surface resistance typically masks deeper fear. Until you diagnose what's genuinely driving pushback and address root causes rather than symptoms, resistance persists regardless of training volume.

This requires creating safe spaces for people to articulate concerns and skilled facilitators who can translate resistance into actionable organizational improvements.

Dimension 6: Cultural Operating System Redesign

Every organization operates on invisible code, the unwritten rules about how things actually function. These invisible rules destroy AI adoption. You must surface them, examine them, and deliberately establish new rules supporting experimentation, learning, and adaptation.

This might mean replacing "don't make mistakes" with "experiment rapidly and share learnings," or shifting from "protect your expertise" to "leverage collective intelligence including AI."

Dimension 7: Professional Identity Evolution

Unlike entrepreneurs accustomed to evolving roles, many employees expect role stability. However, with AI transforming work, identities and purposes are shifting. Helping leaders and team members separate from fixed role identities and reconnect with the impact they create is critical.

This involves transitioning from defining yourself by tasks you perform to defining yourself by problems you solve and value you create. When AI changes the how, your professional worth doesn't vanish; it transforms.

Dimension 8: Ethics, Trust Infrastructure, and Governance

Developing responsible, transparent practices, establishing policies, trust frameworks, and ethical guidelines governing AI usage within teams is vital. People need to know when AI should be deployed and when it shouldn't. Explicit policies build trust in ways vague promises never can.

This includes addressing unofficial AI usage throughout organizations through audits and secure experimentation protocols. Rather than punishing shadow AI use, create transparent channels that encourage open experimentation while reducing hidden risks. Clear governance transforms secretive workarounds into productive innovation.

Dimension 9: Incentive and Performance Alignment

If people are still evaluated and rewarded based on pre-AI metrics and behaviors, they won't adopt new AI-enhanced workflows regardless of training or encouragement. Performance systems must evolve to recognize and reward effective AI collaboration, learning agility, experimentation, and knowledge sharing.

Dimension 10: Regulatory Compliance and Risk Management

Understanding and complying with evolving global AI regulations is essential. New laws govern AI usage in outreach, communications, digital replication, and intellectual property. One of people's primary concerns about advancing initiatives is fear of errors, penalties, or other consequences.

Dimension 11: Stakeholder Coordination

Ensuring everyone, from executives to management to frontline teams to employees to vendors, operates from shared understanding is crucial. Misalignment destroys momentum faster than anything else. When executives are enthusiastic, middle management is doubtful, and frontline employees are terrified, there is no technology that bridges that divide.

Dimension 12: Measurement Frameworks and Feedback Systems

Utilizing adoption scorecards, trust metrics, and feedback mechanisms helps organizations understand how to adopt new practices and build trust. Most companies only measure productivity metrics like tasks completed and time saved. They don't measure trust, psychological security, experimentation willingness, or comfort with ambiguity.

These human metrics matter equally to technical ones because they predict whether AI adoption actually sustains. Regular pulse checks, anonymous feedback channels, and transparent reporting create the visibility needed to course-correct quickly.

How to Actually Build Organizational Readiness

Building stage 4 before rushing toward automation and autonomy starts with honest evaluation. Not a technology assessment, but a trust and readiness assessment.

How psychologically secure is your culture currently? Can people admit mistakes without consequences? Challenge problematic ideas without political repercussions? Ask fundamental questions without being labeled incompetent? If the honest answer is "not really," you have foundational work before AI adoption will sustain.

How aligned is your leadership team? Are executives, managers, and frontline leaders operating from shared understanding of what AI means for the organization? Leadership misalignment destroys momentum faster than any other factor.

What are people genuinely afraid of? Not what you assume they fear, but what they'll share if you create space for honest conversation. Until you address those fears directly, everything else remains a superficial change that won't last.

From there, work happens across all twelve components simultaneously. Start by establishing psychological security so people feel safe voicing concerns and experimenting. Develop leaders who can navigate uncertainty. Create explicit governance guidelines. Reframe roles around impact rather than tasks. Build measurement systems tracking trust and adoption, not just productivity. Address resistance by diagnosing root causes rather than symptoms.

This work lacks glamour and doesn't create exciting vendor demonstrations. But it's the difference between AI adoption that transforms your business and AI adoption that becomes an expensive warning story.

The Decision Every Organization Faces

Throughout history, business success depended on team intelligence. Companies recruited the smartest people and competed based on human cognitive capability.

Now intelligence has become commodified. We have AI systems exceeding human capability in specific domains, and we're approaching artificial general intelligence that will surpass human capability comprehensively.

This terrifies people because their entire career was built on being the most knowledgeable person in the room. Successful companies help people understand that AI amplifies human capability rather than eliminating humans. But that amplification only functions if people willingly engage with AI as collaborative partners.

By this decade's end, two types of businesses will exist. Organizations that constructed stage 4 first, creating sustainable transformation on solid foundations of trust, psychological security, and cultural readiness. And organizations that kept throwing resources at stages 1 to 3 while puzzling over why nothing sustained.

Organizations succeeding at AI adoption don't have superior technology. They have superior cultures where people feel secure enough to experiment, where leaders acknowledge they lack complete answers, and where failure is expected as learning rather than punished as incompetence.

The AI adoption gap isn't a technology problem. It's a leadership, culture, and trust problem. Those problems get resolved with better conversations, better transformation management, and better investment in the human dimension of change.

The journey of AI adoption is fundamentally a human journey. Leadership, culture, trust, and psychological security are the foundation upon which successful AI integration gets built. Companies that comprehend this truth will thrive. AI is a tool for human amplification, not elimination. By investing in people, fostering alignment, building trust, and creating cultures of continuous learning, organizations can unlock AI's genuine potential while ensuring teams feel valued and excited about the future.

About the Author

Scott Hall began his information technology career as an advisor to data technologists of Fortune 500 companies. The advent of the internet and social media led to his first book, *The Blog Ahead—How Citizen Generated Media Is Radically Tilting the Communications Balance*. In the last two decades, he has worked on enterprise search solutions for Forbes Global 2000, and over the last five years he has implemented AI and deep learning models at two of the top five global wealth management firms. In 2024, Scott received his AI implementation, AI consultant, plus data science certifications, which now finds him addressing the needs of SMBs as they embark on a period that will be more impactful than the internet.

And that would be artificial intelligence. Scott's ultimate dream is to apply AI physically to his long-suffering golf swing. Just you wait...

Email: scott@scotthall.com

Website: https://functionalize.com

LinkedIn: https://www.linkedin.com/in/scotthalldotcom/

CHAPTER 10

PRINCIPLED ENGINEERING: GROWING ETHICS WHEN THERE ARE NONE

By Ben Hamilton, PhD
AI Consultant: Tech, Process and Organization
Seattle, Washington

Those who cannot remember the past are condemned to repeat it.

—George Santayana, 1905

Asbestos was good at its job: fireproof, cheap, in everything. Nobody fills a country with a material that doesn't work; the usefulness is real, and the value created the trap. The harm comes slowly, decades between exposure and disease, so the first half of its modern story, "Is this hurting people?" had no proof-grade answer. The absence of the proof gave way, taking until the '70s to be widespread, and litigation turned up the truth: The risks of harm had been already known for decades and suppressed.

The people best placed to see it, did. The signs were there early. There is a rich history, still being written, of people who raised asbestos issues. Most people, however, said nothing. Say nothing. And one reasonable step at a time, we have sixty years of litigation, and asbestos kills tens of thousands of people per year in the United States. Engineering was not the issue. It was the half-second where seeing should have turned into saying, and for most, that's where it stopped.

Certainty was never going to come in time. It still won't. Acting well under uncertainty is the real job.

Why didn't the folks who knew better, speak up? Again, some did. We need a better question: "If you or I were in the same position as one of them, is there any reason we wouldn't do the same?"

The New Principle: Grow Awareness

The engineer who sees something wrong and says nothing rarely thinks of it as a choice. It comes dressed as "not my place," "above (or below) my level," "not my area of expertise," "somebody else already thought this through," "I'm too new," or "I'm too vested." We can mistake this for the things we value: trust, teamwork, ethics, responsibility, efficiency. Biologically, it's the oldest fear we have: being the one who stuck their neck out and got cut.

Here's what I learned about myself: Brains ignore almost everything, on purpose. A filter decides what reaches us and what gets dropped, and it decides by what it already expects to find. Not consciously expect, but subconsciously. What grabs my attention, what slips past unnoticed, what I'd swear wasn't me or them or us—that's the filter, running. Turns out I didn't set it. It's been running on defaults handed to us by habit, by training, by everyone who shaped what we expect. So when the filter skips something uncomfortable and our reason is "outside my expertise," it isn't lying. It's handling the way you feel about yourself, the problem, and the organization so deftly we get comfortable without realizing it's autopilot.

Awareness is catching the filter in the act. Noticing the exact moment you talk yourself out of speaking or get pulled in by language that sidesteps the brain's ancient security system. Learn a category, and perception changes before behavior does. Awareness becomes ethics. With a bit of practice, you build them on purpose.

Language

The right words bypass the filters and make it so we don't have to think: pre-packaged perceptions get waved through because of the novelty, authority, or identity they carry. Here are four examples I see often, but there are many more.

Marketing Language: It isn't trying to inform you. It's writing a world and casting you in it, a version of you that needs the thing, feels the lack, completes the story by buying in. The information is incidental; your role is the point. The tell: It describes who you'd be, not what the thing is. When a sentence is busier with your identity than with its own subject, you're reading marketing, whether it's selling a product, a company, or a future for AI. Look for pivots, too, after push back. "Fifty percent of white-collar jobs will be replaced by AI" is not an engineering guarantee. It's a marketing strategy. The roll back was clockwork. The cycle continues.

Language about the Law: In engineering, the law is the floor. Not the ceiling. We still need it, but we need to understand what legal language means for our systems and our customers. It's slow and backward-looking, built on systems that move far slower than the technology they're trying to govern, so "legal" and "ethical" drifted apart a long time ago. The gap widens every year. Watch what people do with that gap. Federal and state laws are about criminal laws. They have tough bars. Civil courts are about how we harm each other. The tell: "We are within the law" when a place has an army of engineers and scientists who can actually do better. "We admit no wrongdoing" when civic payouts are huge. Pretending civic courts aren't important because it's money and not crime.

Accountability Language: The word is a substitute for the act. "I take full accountability" or "I'm a high-accountability person" arrives sounding responsible, defuses the harder question, buys compliance, and obligates the speaker to nothing. Real accountability isn't the sentence; it's what changes after, and you can only see it in the past tense. The tell: The language arrives without transparency, without a plan, without anything different on the other side of it. Pay attention to the results. See it in leaders, in our organizations, and most critically in ourselves.

AI Language: The biggest one, and the emptiest. "AI" can mean a model, a feature, a vendor, a layoff, a marketing strategy, a technology strategy, the end of the world, AGI, or the path to utopia. Sometimes at the same time. This fog is confounded by market hype, but it contains uses that are both beneficial and not. And that gives us an opportunity to use AI to envision something new, gives us a stand-in for a person who isn't taking responsibility, or is a reason to cut jobs, and more. The tell: AI language needs to be swapped for a more specific thing, preferably measurable. If that's missing, a plan is missing, too. Don't take their word for it.

The rule underneath them: Impeach the one-word answer. The two-word phrase. The label that arrives where a thought should be, and a thought that takes the place of experience. It extends past these four to all of it: slogans, mission statements, witticisms, conventional wisdom, the adage that ends an argument before it starts. Each of these is a shortcut, a pre-packaged perception you accepted without checking. These shortcuts, and our comfort with them, are inherited and often too shallow.

Histories Become Futures

Santayana's line is the most quoted about history. It's an adage standing where a method should be. It feels like wisdom and tells you nothing to do. And it points at the wrong faculty. "Remember the past," as if the past that runs you were ever yours to witness. The histories that decide what your filter shows you and whether you flinch were installed before you were aware. You *can't* remember them, as he demands. You have to discover them.

Three of them, oldest to newest, and you've already met the first. The oldest is the flinch: 200,000-year-old default that can't tell a threat of layoff from exile from the clan, so it handles them the same. The next was set in childhood, in rooms you probably don't remember: what behavior got attention, who gets to speak and when, whether being wrong is welcomed, whether the truth gets punished. You might think today that "I am not that kind of person" is about identity and truth, when it comes in large part from childhood, which is also solving its part in 200,000-year-old problems.

The third is the newest: the organization. An organization is a few thousand inherited defaults wearing a logo, and it conceals those defaults and their impacts out of sight of the people in charge. You can at least feel your own flinch. An organization can't feel anything; it just runs its scripts and calls the output reality.

None of these were yours to choose. They were set before you arrived, by genes and parents and organizations that set up your filter. Santayana's words don't capture that the work was never "remembering," because nobody knew this a century ago. Our job now is becoming aware of history and learning to aim ourselves on purpose.

Abundance and Scarcity

The whole chapter has been working against you, surfacing threats, cheap words, the flinch, the defaults you never chose. Engineers are amazing at scarcity. It's near the whole job. Details, scrutiny, build fast, design slow, get feedback, compare constantly, ruthless precision. When our filters are trained for this, that's what we see when we look around. Opportunities that won't work, relationships that feel too hard to start, knowing we're not that great at speaking up. Some of these run deep and strong. And when the world gives us feedback that supports the scarce assessment, we were right all along.

Turns out, brains come with some other states. One of them is abundance. "Expect good things and you'll find them" sounds like somebody's bumper sticker, not a way to solve hard problems.

Abundance is a real human thing, and if it is around for us now, it's been around since we still lived in tribes. Confront challenges like they're going to end up alright. Like the possibilities will unfold in a way you can handle. That's the wedge. No acausal magic required. No change in how the world works. Get your filters working *for* you, so they can start seeing solutions to problems that scarcity doesn't.

Here's the language the bumper sticker gets wrong. You don't re-aim the filter by expecting to. If wanting it were enough, you'd have wanted your way out of the flinch years ago. You re-aim it by acting: once, small, before you feel ready. Say the thing in the meeting. Send the email you've been drafting in your head. Put one sentence on the page. The rep is tiny on purpose; the size is never the point, the direction is. You're showing the filter what to start looking for, and it learns from what you did, not what you intended.

It won't feel good. The flinch is still there, the same ancient forecast that you'll be cast out for it. You do it anyway, nothing catastrophic happens, and the filter logs that. Then you do it again. It never gets easy, exactly. What changes is that you stop waiting to feel ready, because you've collected enough evidence that ready was never coming and the step worked anyway. That evidence, worn on the outside, is what people read as confidence. It isn't confidence. Confidence is making the reps while it still feels bad. The look is what's left over – a symptom of the work itself.

Now for the engineer's dichotomy: We don't get to turn scarcity off. You can't, and you shouldn't want to. Our adversarial minds hunt the failure case, distrust the happy path, assume the inputs are the thing nobody wants to talk about, and see piercing truths. Scarcity is how the work gets done, it's how technology goes from imagination to production. The trouble is when it leaks off the problem and onto the people. Defending turf, reading every hard call as a threat to your standing. It's keeping scarcity pointed at the work and growing the skill to flip modes of operation. We're a wildly varied group, and there are words now for things we didn't have before. Some of us don't admit it, but we're a person, too - nearly all of our challenges have labels, most have answers. Some have paths through.

AI has been the next hard thing for a long time. The current wave is new, genuinely powerful, and the early signals of harm are the kind that got pushed off and suppressed for sixty years with asbestos. Fearing AI and idolizing AI are two major trends right now, but we, the engineers, need to be aware of things we have never needed to know before. We need tools we've never needed before, tools that reach past the model itself. Our engineering shapes the physical world, leaving a footprint on our environment, on people, on jobs, and on society. Some of our engineering has a psychological footprint in ways only now we are able to fathom. Before the ethics have settled. Before the litigation has completed. We have to build.

The challenges scale, as do the filters. The filters extend to all of us, not just engineers. It's toughest at the top, because the structure hides the organization's defaults from the person most responsible for them, and seeing past it is often impossible. The flinch is the same human one, with what or why being very different from the engineers. The difference is up there you're handed the most accountability and the least visibility, by design. Even so, growing awareness of how our histories, fields, and language come together can reveal where we can be better positioned today than ever before, to be conscientious, ambitious, innovative, and successful. Feedback in either direction has greater depth, lessons are learned by individuals and organizations they would never have otherwise, tiny steps that feel like big leaps are taken more often, and the cycle of trust and understanding fuel organizational growth.

About the Author

Ben Hamilton was native to the Midwest United States, and earned a degree in chemical engineering from Purdue University, followed by a PhD from the University of Minnesota. His early work was in pharma, nanomaterials, and chemical processes. By 2009 he was automating his own job, building simulations and models, and analyzing things everybody took for granted. He spent several years as a process engineer consultant at the height of the fracking boom and began global strategy analysis and writing about corporate culture for global leaders. In 2017, he combined his process engineering and simulation expertise with data science, where dedication to decision science, product, and risk engineering translated to helping startups get funding, impacting across eight industries, and leading product innovation teams and responsible AI initiatives at a global consultancy.

He made enough mistakes to learn voraciously from others and possesses a balance of important lessons and impact as high as nine figures and as many as 6,000 people. AI is the one where the biggest impacts come from navigating mistakes and protecting from waste.

Ben operates as a consultant through BDHC (est. 2019), helping leaders solve for and measure business outcomes when AI is both so promising and so hyped. Every engagement looks different, but what time has shown, is when you come in to partner on delivering outcomes, AI value ends up 1% AI, 19% tech, and 80% people process and structure. He partners with leaders to solve the whole problem. He shares with non-engineers the engineering lessons nobody should be expected to know and mentors folks of different neurotypes and health conditions.

Email: ben@bdhamilton-consulting.com

Website: www.bdhamilton-consulting.com

LinkedIn: https://www.linkedin.com/in/bdhamilton

CHAPTER 11

FROM SAAS TO AGENTS AS A SERVICE: HOW AI IS REWRITING THE RULES OF CHANNEL PARTNERSHIPS

By Latif Hamlani
Founder of Partner Assistant AI
San Francisco, California

The best way to predict the future is to invent it.

—Alan Kay

For decades, the technology industry has relied on a familiar equation: Build software, sell licenses or subscriptions, and distribute through a carefully managed network of partners. This model—while effective—was built for a slower, more linear world. Today, that world no longer exists. The pace of innovation has accelerated, ecosystems have exploded in complexity, and the volume of data flowing through channel organizations has grown beyond what any human team can reasonably manage.

We are now at an inflection point. Artificial intelligence is not simply improving existing processes; it is fundamentally reshaping how software is built, sold, and delivered. The next major shift is already underway: a transition from software as a service (SaaS) to agent as a service, where intelligent AI agents actively participate in and optimize the entire channel and partnership lifecycle.

This chapter explores how AI is transforming channel partnerships and software distribution, why traditional models are breaking down, and how agentic AI represents a new operating system for the partner ecosystem.

The IT Market Reality: A $5.6 Trillion Ecosystem Under Pressure

To understand why this transformation matters, we need to start with the scale of the market. The global IT market is valued at approximately $5.6 trillion, with services representing the largest share, followed by software. This alone should reshape how we think about distribution. Software does not exist in isolation; it is bundled, implemented, customized, secured, and supported through services, most of which are delivered by partners.

Channels are not a side strategy. They are the strategy. Data shows that over 70% of software sales involve channels, and an astonishing 96% of sales are partner-assisted in some form. Even so-called "direct" sales are rarely truly direct. Partners influence demand, shape customer decisions, accelerate adoption, and determine long-term success.

Yet despite their importance, partner ecosystems remain one of the most under-optimized parts of the enterprise. Channel leaders are expected to manage thousands—or even tens of thousands—of partners across ISVs, systems integrators (SIs), managed service providers (MSPs), resellers, and distributors. Each partner brings its own capabilities, incentives, relationships, and performance history.[1] The resulting complexity is staggering.

From Licensing to Subscriptions to Marketplaces

The software distribution model has already gone through several major evolutions. First came perpetual licensing, where software was sold as a product and upgraded periodically. Then came subscription-based SaaS, which shifted revenue to recurring models and emphasized continuous value delivery.

Today, we are in the midst of another shift: Cloud marketplaces have become a dominant route to market. Hyperscalers like AWS, Microsoft, and Google have turned marketplaces into powerful distribution engines, bundling procurement, billing, and deployment into a single workflow.

Marketplaces have lowered friction, but they have also introduced new challenges. Vendors now compete in crowded digital shelves. Partners must navigate complex incentive structures. Channel managers must align marketplace motions with traditional partner programs.

What was once a linear value chain has become a dense, multi-directional network.

The Hidden Cost of Ecosystem Complexity

As ecosystems scale, the limitations of traditional channel management become painfully clear. Most organizations still rely on a patchwork of CRM systems, partner portals, spreadsheets, emails, PDFs, and tribal knowledge. Critical information lives in unstructured formats: meeting notes, call transcripts, enablement decks, Slack threads, and personal inboxes.

This creates several systemic problems:

- **Decision Latency:** By the time insights are surfaced, opportunities have already passed.

- **Partner Misalignment:** The wrong partners are often engaged for the wrong deals.
- **Revenue Leakage:** At-risk accounts go unnoticed until churn occurs.
- **Burnout:** Channel managers spend the majority of their time on administrative tasks instead of strategic work.

Ironically, as partner ecosystems become more important, humans are becoming the bottleneck.

Why More Headcount Isn't the Answer

Historically, the solution to complexity has been to add more people. More partner managers. More operations staff. More analysts. But this approach does not scale. Each additional layer increases coordination costs and slows execution.

The real issue is not effort. It is cognitive load.

No human, no matter how experienced, can simultaneously track thousands of partner relationships, monitor real-time performance signals, analyze historical outcomes, and make optimal recommendations at scale. This is precisely the type of problem AI is designed to solve.

The Emergence of Agentic AI

This brings us to the next evolution: agent as a service. Unlike traditional software tools that require users to query dashboards or pull reports, agentic AI operates proactively. These agents can ingest massive volumes of structured and unstructured data, reason across it, and take action or recommend action in real time.

In the context of channel partnerships, an AI agent can function as a tireless, always-on partner strategist.

Imagine an agent that:

- Continuously monitors customer and partner health signals
- Identifies accounts at risk before humans notice
- Understands partner strengths, certifications, and past performance
- Recommends the best partner for a specific opportunity
- Learns from outcomes and improves over time

This is not automation for automation's sake. It is augmentation, freeing humans to focus on what they do best: relationships, creativity, and strategic thinking.

From Reactive to Proactive Partner Management

One of the most powerful shifts enabled by AI is the move from reactive to proactive ecosystem management. Traditionally, channel teams respond after something goes wrong: a deal stalls, a partner disengages, a customer churns. By then, the cost of intervention is high.

Agentic AI flips this model. By analyzing patterns across accounts, partners, and engagements, AI can surface early warning signals and opportunity indicators. Instead of asking, "What happened?" leaders can ask, "What should we do next?" This fundamentally changes the role of the channel manager—from administrator to orchestrator.

Inside the Partner Assistant Concept

At Partner Assistant.ai, the goal is to build exactly this type of intelligence into the core of partner ecosystems. In early wireframe demonstrations, the AI agent analyzes customer accounts and flags those that may be at risk based on usage patterns, engagement signals, and historical data. It then recommends specific partners who are most likely to succeed with that customer based on existing relationships, technical expertise, industry focus, and past win rates.

What makes this approach powerful is not just speed, but contextual understanding. The agent does not treat partners as interchangeable. It recognizes that ecosystems are built on trust, history, and nuance, factors that traditional systems struggle to model.

The Human Still Matters More Than Ever

A common fear around AI is that it will replace people. In reality, the opposite is happening. As AI takes over repetitive and analytical tasks, the human aspects of partnerships become more valuable, not less.

Trust cannot be automated. Empathy cannot be templated. Strategic alignment cannot be reduced to a rule set. Agentic AI creates space for humans to do the work that actually moves ecosystems forward: building long-term relationships, co-innovating solutions, and aligning partners around shared outcomes.

Designing AI With, Not For, Customers

One of the most important lessons in building AI for the enterprise is that it cannot be developed in isolation. Channel ecosystems are deeply contextual, and every organization operates differently.

That is why Partner Assistant.ai is taking a design-partner approach, co-developing the agent with early customers. This customer-obsessed model ensures that the AI reflects real-world workflows, not theoretical ones. AI is most powerful when it is shaped by the people who will rely on it every day.

A New Operating Model for the Channel

We are witnessing the emergence of a new operating model for software distribution, one where intelligence is embedded directly into the fabric of partnerships. In this model:

- Data is no longer passive; it is actionable.
- Insights are delivered in real time, not after the fact.
- Partners are matched strategically, not opportunistically.
- AI agents work alongside humans as trusted advisors.

This is not a distant future. It is happening now.

Conclusion: Inventing the Future of Partnerships

The transition from SaaS to Agents as a Service represents more than a technological upgrade. It marks a philosophical shift in how we think about scale, intelligence, and collaboration.

Channel ecosystems are too complex, too dynamic, and too important to be managed with yesterday's tools. Agentic AI offers a path forward, one that amplifies human capability rather than replacing it.

The future of software distribution will belong to organizations that embrace this shift early, design thoughtfully, and remain relentlessly focused on partnership outcomes.

As Alan Kay famously said, the best way to predict the future is to invent it. In the world of channels and partnerships, that invention has already begun.

References

1. Canalys, now Omdia, "Channels & Partner Ecosystems," accessed June 30, 2026, https://omdia.tech.informa.com/advance-your-business/channel.

About the Author

Latif Hamlani is the CEO and founder of Partner Assistant.ai, an AI-native platform focused on transforming how enterprise software companies manage and scale their partner ecosystems. With deep experience across cloud marketplaces, channel strategy, and enterprise go-to-market models, Latif works at the intersection of AI, partnerships, and software distribution. He is a frequent speaker on the future of software ecosystems and agentic AI, and is passionate about building technology that augments human relationships rather than replacing them.

LinkedIn: https://www.linkedin.com/in/latifhamlani/

CHAPTER 12

BEFORE FIRE, BEFORE THE WHEEL, BEFORE AI, THERE WAS THIS

By Michael Harris
Founder, Storibot.ai | Narrative Intelligence Engineer
Calgary, Canada | San Francisco, California

The most powerful person in the world is the storyteller.
—Steve Jobs

Jobs spoke these words while he was running Pixar in 1994. He didn't deliver this statement during a high-stakes Apple keynote or while launching a world-changing gadget. He said it at a film studio that he helped build, surrounded by animators and storytellers, people who spent their days figuring out how to make you cry over a cartoon fish.

He wasn't done. He added: "The storyteller sets the vision, values, and agenda of an entire generation that is to come."

An entire generation yet to come. Not the people in the room. Not the shareholders. The people who haven't been born yet. That is the true reach of a story. It crosses borders. It crosses centuries.

Now let me take you somewhere. Far from a film studio or conference room, miles away from a theater, nowhere near a TED stage. Let's go back to the beginning. To a cave.

One Hundred Thousand Years Ago, Someone Had Something to Say

The cave is called Blombos Cave. It sits on the southern coast of South Africa, overlooking the Indian Ocean. The wind has been blowing off that water for a very long time. And in the layers of sediment inside that cave, archaeologists stumbled upon a find that forced a total rewrite of our evolutionary timeline. Ochre. Red ochre, deliberately engraved with geometric patterns. Cross-hatched lines. Abstract designs. Scored into stone with intention, with care, with the specific and unmistakable mark of a mind that wanted to make something mean something. The sediment around these pieces has been dated between 75,000 and 100,000 years ago.

One hundred thousand years ago, a human being picked up a piece of red stone and made a mark that was meant to communicate. Meant to be seen. Meant to say, "I was here. This means something. Remember this. This is my story."

That is the beginning of storytelling. Before Shakespeare, before the printing press, and before film crews could capture anything at all. A smear of red ochre on a rock face in a cave by the Indian Ocean, made by someone whose name we will never know, whose language has been gone for longer than civilization has existed.

And yet here we are. Still doing exactly what that person did. Still making marks. Still trying to say, "This means something."

Then Came Fire, and Everything Changed

About 400,000 years ago, we figured out fire. I know what you're thinking. Fire is about warmth. Fire is about cooking. And yes, absolutely. Richard Wrangham has made a compelling case that cooking food is what expanded the human brain, because cooked calories are more efficiently absorbed, and a bigger brain needs more fuel. So, in a very real sense, we barbecued our way to intelligence.

But fire did something else. Something nobody talks about nearly enough. Fire gave us night.

Before fire, when the sun went down, you went to sleep. There was nothing else to do. The darkness was full of things that ate you, and your best option was unconsciousness. But with fire, you got four, five, six more hours every evening. Hours in which you couldn't hunt, couldn't forage, couldn't do any of the survival work that filled your days. You just sat there. Together. Warm. Looking at the flames.

And so, you told stories.

University of Utah anthropologist Polly Wiessner spent decades studying the Ju/'hoansi Bushmen of Namibia and Botswana. She analyzed 174 day and night conversations recorded across decades of fieldwork. What she found was simple and extraordinary. During the day, people talked about practical things. Food. Social matters. Who was doing what to whom. Standard human business. But at night, around the fire, more than 80% of conversation became stories. Tales of real people. Folktales. Myths. Encounters with the spirit world. And it was all wrapped in music and gesture and the kind of full-body performance that you don't get in a Zoom call.

The campfire was the first theater. The first broadcast network. The first content platform. And it ran for hundreds of thousands of years before anyone thought to write any of it down.

The Stories That Remembered the Ocean

Here's my favorite story about stories.

The Aboriginal Australians arrived on their continent at least 65,000 years ago. They brought with them a system of storytelling called songlines. The Dreamtime. Stories in which ancestral beings traveled across the landscape at the beginning of the world and sang everything into existence. Every mountain. Every waterhole. Every creature. The world was not built. It was sung.

The paths those ancestral beings walked became the songlines. Vast, interlocking networks of narrative that crisscross the entire Australian continent, some stretching for thousands of kilometers across multiple language groups and cultural nations. By learning the songs in the correct sequence, a person can navigate terrain they have never seen. The song is the map. The story is the GPS.

Recently, marine geologist Mick O'Leary identified that certain songline narratives accurately describe underwater watering holes. Deep blue watering holes that are now fourteen meters below sea level. Below the ocean. Because when those songlines were composed, sea levels were lower, and that land was above water. The songlines were composed more than 7,000 years ago and have been passed down, generation to generation, through song and ceremony, without a single word written down. And they are still accurate.

An elder recognized features in a songline he had known all his life. Features he had never seen. Features that are now at the bottom of the sea.

That is entertainment and infrastructure. That is the most durable memory our kind has ever built. Story encoded into ceremony, passed through the human voice and body, outlasting empires, outlasting languages, outlasting the shoreline itself.

If you are ever tempted to think of storytelling as soft, as a nice-to-have, as the decorative layer on top of something more serious, I want you to remember the songlines. I want you to remember that story about the deep blue water and the shape of the ocean floor.

The Oldest Story We Can See

In 2024, researchers announced that a cave painting in Sulawesi, Indonesia, had been definitively dated to at least 51,200 years old. It shows three human-like figures interacting with a wild pig. One figure extends both arms and appears to hold a rod. Another figure presses something toward the pig's throat. A third is upside down, reaching for the pig's head. Whatever it depicts, it is unmistakably a scene. Multiple characters. A shared action. A moment with a before and an after.

The researchers called it, without any overstatement, the oldest known example of visual storytelling in the world.

Fifty-one thousand years ago. And someone understood that you need characters. You need action. You need something happening. The bones of narrative were already there, fully formed, in a limestone cave in Indonesia, when most of Europe was still covered in ice.

We didn't invent this. We inherited it. All of us. Every person who has ever written a story or screenplay, drafted a brand manifesto, told a joke at a dinner table, tried to explain to their kid why the dog had to go to the vet and wasn't coming back. We are the latest participants in a tradition that is older than any other human technology. Older than agriculture. Older than the wheel. Older than the city. Older than writing by so many tens of thousands of years that writing barely shows up on the timeline.

And Then We Let the Machine Talk

I have spent thirty years in media and entertainment. I have worked on projects with budgets larger than some national parks. I have sat in writers' rooms at two in the morning arguing about why a character would make a decision that nobody in the audience would make, because that's the decision that would reveal who the character was. I have studied Alfred Hitchcock, Stanley Kubrick, Steven Spielberg, and Quentin Tarantino to find the single line in a scene that unlocks everything else. I know what story looks like when it's working.

And I know what it looks like when it isn't.

Right now, the internet is drowning in content that has no story in it. AI-generated copy that informs, technically, while connecting to nothing. Text that checks the box without hitting the nerve. Blog posts that contain all the correct information and land with the emotional impact of a warranty document. People are producing more words than at any point in human history, and fewer of them are doing anything to anybody. I've given a name to what you get when volume beats intention. It's called "bionic slop."

And we're drowning in it.

This is not AI's fault. AI is a tool, and a remarkable one. The fault is in how we are using it. We handed the machine the ability to produce language, and we forgot to also hand it the 100,000 years of accumulated human understanding about what language is for. That is why I created a Narrative Intelligence platform that teaches machines the craft of story, not merely the vocabulary of language.

Language is not for information transfer. Language is for connection. For meaning-making. For the specific experience of sitting by a fire, or a screen, or a page, and feeling less alone in the universe because someone else has articulated something true about being alive. That is what story does. That is what story has always done. That is what no algorithm, on its own, knows how to do.

Who's Driving the AI Boat

Every generation faces its version of this problem. Every time a new technology arrives, the question is the same: Who is driving? The printing press could print the Bible or it could print propaganda. Radio could broadcast Beethoven or it could broadcast a dictator. Television could show you the moon landing or it could sell you cigarettes. The technology does not make the choice. People make the choice.

We are the people who must make the choice right now. Today. With this technology. Because AI is not going to slow down. It is not going to get less capable. It is not going to become less central to how the world communicates. The question is not whether AI is going to shape the stories of the next generation. It is already shaping them. The question is whether the people who understand storytelling are going to be the ones shaping how AI tells them.

I think about the person at Blombos Cave. The one who picked up the ochre. They did not have to make that mark. They had food. They had fire. They had shelter. There was no practical reason to make that mark. And they made it anyway. Because something in them understood, at a level deeper than strategy, that meaning had to be preserved and shared. That experience had to be transmitted. That what it felt like to be human in that moment had to be passed forward.

We are that person right now. In this specific moment, with this specific technology. The ochre is in our hands. So is the AI.

The Oldest Technology Has to Survive the Newest One

"The storyteller sets the vision, values, and agenda of an entire generation that is to come." Jobs was right. He was always right about the things that mattered, which is annoyingly beautiful, but there it is.

The generation yet to come is going to grow up in a world where AI produces a significant portion of what they read, watch, and hear. They are going to be surrounded by machine-made content from the moment they are old enough to look at a screen. The question of whether that content carries genuine story, human truth, and a shape that means something is not an aesthetic question. It is not a philosophical question for a seminar room. It is the most practical question imaginable: What kind of human beings is the next generation going to become?

Stories build us. They always have. From the ochre engravings at Blombos Cave to the paintings in Sulawesi to the songlines that remembered the deep blue sea floor, the record is clear: We are shaped by our narratives. We think in story. We bond through story. We make sense of suffering through story. We become the people we are going to be because of the stories we are told about who we could be.

If the stories of the next generation are hollow, the next generation is going to have to work very hard to fill themselves back up. And that is a cost I am not willing to accept without a fight.

This story is that fight. Not against AI. With it. Alongside it. Steering it. The goal is not to go back to the cave. The goal is to bring the cave with us. To carry forward the understanding that has made storytelling the most durable technology in human history, and to make sure that understanding is baked into the way we use AI.

Now we tell the right stories.

About the Author

Michael Harris has spent over thirty years in the writers' rooms, sound stages, and story conferences where Hollywood does its real work: the unglamorous, two-in-the-morning business of figuring out why a character does what they do and why it matters. He's collaborated with visionaries, creators, and believers. He has built and exited a successful production studio, and absorbed, frame by frame, what separates a story that moves people from content that merely informs.

But somewhere along the way, the campfire became dimly lit. The internet had filled up with words that checked every box and touched nothing. What Michael calls "bionic slop" has polluted the information highway with meaningless content. So, he did what storytellers have always done when the world needs a new myth: He picked up the modern equivalent of the ochre stone.

Today, as founder of Storibot.ai, Michael is teaching artificial intelligence the 100,000-year-old craft of meaning-making, fusing the storytelling frameworks into technology built to put soul back into how businesses, brands, and people communicate. Michael calls this "Narrative Intelligence."

He's not anti-AI. He's pro-human. His mission is to make sure the most powerful storytelling tool ever built doesn't forget what story is *for*: connection, not just information.

Off the page, Michael coaches and mentors MBA students, grad students, and high school students, as well as youth fastpitch softball and hockey, where he teaches the same lesson he writes about: that everyone is already the hero of something, they just need someone in their corner who believes it.

Because here's the truth Michael wants you to walk away with: This was never just an essay about cave paintings and campfires. It's an invitation. You are the one holding the ochre now. Your story in your business, your relationships, your own unrepeatable life is yours to engrave. Michael's not here to tell it for you. He's here to remind you it's worth telling and to hand you back the stone.

Email: info@storibot.ai

Website: https://www.storibot.ai

LinkedIn: www.linkedin.com/in/michael-harris-725abb1

CHAPTER 13

TRUST AS INFRASTRUCTURE: WHY PROACTIVE AI GOVERNANCE IS YOUR FASTEST PATH TO SCALE

By Dr. Sheik M. Isaacs
AI Governance & Enterprise Data Executive
Tampa, Florida

When trust goes up, speed goes up and cost goes down. When trust goes down, speed goes down and cost goes up.

—Stephen Covey

Imagine your organization has just deployed a sophisticated AI system to approve mortgage applications. The model is brilliant, processing thousands of applications in seconds, with accuracy rates no human team could match.

Then, one morning, a regulator calls. "Can you explain precisely why your system denied this applicant?"

Silence. That silence—the inability to answer—is the most dangerous place a modern enterprise can find itself. And it is far more common than most leaders realize.

Advances in computing power gave companies powerful new tools: systems that predict customer behavior, algorithms that approve loans in seconds, and models that analyze data that would take humans years to process. In banking, healthcare, and insurance, the excitement was immediate. But as these systems move from controlled experiments into core production workflows, a stark reality is setting in: The real bottleneck to scaling AI is not a shortage of technical capability. It is a profound deficit of trust.

When an AI model makes a multi-million-dollar credit decision, recommends a patient's treatment plan, or executes a complex market trade, three questions emerge and they are not technical:

1. Should the system have made that choice?
2. What assumptions governed the outcome?
3. Can we prove its validity to a regulator, a board of directors, or an affected consumer?

These are not rhetorical questions. Regulators are asking them now. Boards are asking them now. Consumers are beginning to ask them too. This chapter answers each one directly through a practical, three-pillar framework, and the conclusion maps exactly where each answer lives.

Why Traditional Governance Fails AI

Historically, regulated industries managed new technologies through defensive compliance: retrospective audits, checklist approvals, and siloed gatekeeping. That approach worked when software followed predictable, linear rules. You could trace a decision to its source and audit the result with confidence.

AI systems are fundamentally different. They are dynamic. They learn, adapt, and change over time based on the data they consume.

Applying legacy governance to AI creates a dangerous paradox: Apply it too strictly and valuable models get trapped in approval purgatory, destroying business value; apply it too loosely and teams bypass safety protocols, exposing the organization to regulatory penalties, bias claims, and reputational damage that takes years to repair.

The table below captures why the rules written for traditional software do not hold when the software learns, adapts, and evolves on its own.

Traditional Software	AI/Machine Learning Systems
Static, rule-based code	Dynamic, data-driven models
Predictable, traceable outputs	Probabilistic, evolving outputs
Retrospective audits: effective	Retrospective audits: insufficient

Figure 1: Why legacy governance frameworks fail when applied to AI.

Think of it like a high-performance racing car. Superior brakes do not slow the driver down. They allow safe acceleration into corners that would otherwise be impossible. Robust AI governance does not block innovation. It is the very mechanism that enables it.

The Three Pillars of Trust as Infrastructure

Transforming governance from concept to enterprise infrastructure requires a practical framework. That framework rests on three interconnected pillars, each directly answering one of the three questions this chapter opened with.

Pillar 1: Knowing Your Data: Provenance, Lineage, and Quality
Answers: "What assumptions governed the outcome?"

Every AI model is only as credible as the data that trains it. And data is never neutral. Every dataset carries assumptions, assumptions about which historical patterns to repeat, whose outcomes to optimize, and what "success" looks like. Those hidden assumptions become the invisible rules governing every decision the model makes.

Data provenance—the fully documented history of data origins, transformations, and movements—makes those assumptions visible and correctable. Data lineage maps the complete journey of that data, every transformation, system, and handoff from its original source to the model. Data quality ensures that what arrives at the model meets the standards required to produce reliable, unbiased outputs. In practice, this requires automated lineage pipelines that track how data is collected and processed, continuous quality monitoring as a risk discipline, formal data policies that define bias-screening thresholds, and clearly defined ownership roles: data stewards managing day-to-day quality and data owners accountable for legal permissibility and risk.

Real-World Example: The Automated Recruiting Bias

A large enterprise trained an AI recruiting tool on a decade of historical hiring data. Because tech roles had historically skewed male, the model learned an unstated assumption: Male career patterns equal success. It began systematically penalizing qualified female applicants, not by design, but because nobody examined the assumptions buried in the training data. Without provenance controls and representativeness checks, the system scaled decades of human bias in milliseconds. What assumptions are hidden in your training data?

Pillar 2: Algorithmic Explainability and Model Transparency
Answers: "Can we prove its validity to a regulator, a board of directors, or an affected consumer?"

The era of the "black box" AI system is over. It is no longer acceptable to respond to scrutiny with "the model is too complex to explain." Proving validity is now a legal obligation. Under Europe's AI Act, deploying high-risk AI systems without adequate transparency can result in fines of up to €15 million or 3% of global annual turnover, whichever is higher.

Consider a retail bank that denies a long-time customer a routine car loan. Consumer protection law requires the bank to explain why. If the risk team cannot isolate the reason, the result is regulatory sanctions, reputational damage, and a lost customer. Explainability converts algorithmic logic into human-readable evidence, the kind that holds up to a regulator, reassures a board, and respects the consumer's right to understand a decision that affects their life.

What constitutes adequate proof varies significantly depending on who is asking and what is at stake for each audience. The breakdown below maps those differences:

Stakeholder	What They Need to Know	Why It Matters
Regulators	Mathematical proof that the model does not rely on variables violating fair-lending or anti-discrimination law	Legal and licensing compliance
Board of Directors	Confidence that AI decisions align with corporate strategy, risk appetite, and fiduciary duty	Executive accountability and investor confidence

Operators	Why a specific recommendation was generated, so human judgment can be applied before action	Operational accuracy and accountability
Consumers	A plain-language explanation of the factors that influenced their outcome	Consumer rights and trust

Figure 2: Explainability requirements by stakeholder.

Explainability cannot be retrofitted after launch. It must be designed into the model's architecture from day one.

Pillar 3: Continuous Guardrails and Algorithmic Auditing Answers: "Should the system have made that choice?"

Traditional software is static. AI models are not. Because they change over time, the question "Should the system have made that choice?" cannot be answered once at launch and forgotten. It must be answered continuously. This requires continuous guardrails—real-time automated checkpoints that monitor running models and route decisions to human review when thresholds are breached—and algorithmic auditing, which is the structured, ongoing evaluation of a model's fairness, accuracy, and compliance.

Without these guardrails, models degrade silently through three vectors: concept drift, where the relationship between inputs and outputs shifts (e.g., a fraud model trained on pre-pandemic spending begins flagging normal e-commerce as suspicious); data drift, where live data no longer resembles the training data (e.g., a healthcare AI trained on urban elderly patients fails in a rural clinic serving a younger demographic); and model drift, the measurable consequence of both, thus eroding accuracy and reliability while the system continues running in production, often invisibly.

The framework below maps each layer of those guardrails in practice: what it monitors, what triggers a response, and what the system does next.

Guardrail Layer	What It Monitors	Trigger Condition	Response
Performance Monitoring	Accuracy and prediction drift	Metrics fall below preset thresholds	Alert + model review
Bias Detection	Disparate impact across groups	Fairness metrics exceeded	Suspend + audit
Data Anomaly Detection	Quality failures in incoming data	Every data ingestion cycle	Flag + remediate
Human-in-the-Loop	High-stakes decisions	Model confidence below threshold	Route to human reviewer

Figure 3: The Continuous Guardrail Framework.

In Practice: Guardrails at Work

An insurance firm using an ethical MLOps framework co-designs model boundaries from day one, banning zip codes as proxy variables and coding automated bias limits into the deployment pipeline. When a regional economic shift later alters local repair costs, a continuous guardrail triggers a performance alert, reroutes borderline claims to human adjusters, and flags the data team to retrain the model. Compliance maintained. Operations uninterrupted. Zero downtime.

The Cultural and Leadership Shifts of Trusted AI

The three pillars are not purely technical challenges. Implementing them requires a cultural transformation in terms of rewriting how multi-disciplinary teams collaborate. Traditional technology projects follow a sequential pattern: data engineers build, data scientists model, engineers deploy, and compliance audits the finished product. By the time compliance reviews, governance failures may already be baked in, requiring the model to be scrapped at enormous cost.

The modern approach is ethical MLOps, a cross-functional model where governance is embedded throughout development, not applied at the end. The shift from one model to the other changes not just how teams work, but how quickly and safely AI reaches production. The contrast between the two approaches is captured below:

Traditional Sequential Model	Ethical MLOps Model
Compliance reviews the finished product	Compliance co-designs from day one
Governance is a gate at the end	Governance is embedded throughout
Bias issues discovered late	Ethical boundaries defined before development
Speed and safety in tension	**Speed and safety in alignment**

Figure 4: From sequential silos to ethical MLOps.

Critically, the cross-functional team must define—in writing, before a line of code is written—which decisions the model is authorized to make autonomously and which must always involve a human. This is how the question "Should it?" gets answered before a problem occurs, not after.

This cultural shift also demands redefining what model success looks like. For years, performance was measured on accuracy, precision, and recall alone. In a governed framework, those metrics must be weighted equally against governance metrics:

- A model with 95% accuracy that cannot be explained or audited is worth zero to an enterprise.
- A model with 91% accuracy, flawless provenance, and real-time bias guardrails is a valuable, scalable asset.

Governance as Competitive Advantage

The immediate motivation for building ethical AI infrastructure is risk mitigation. The long-term payoff is strategic dominance. Organizations with standardized governance pipelines move AI from design to production faster than competitors who treat compliance as an obstacle. Safety becomes the catalyst for speed, not its enemy. As consumer awareness of algorithmic bias grows, the ability to demonstrate transparent, ethical AI becomes a powerful brand differentiator. And as regulatory pressure increases globally—Europe's AI Act, evolving US federal mandates, SEC and Federal Reserve guidelines—organizations built on adaptable trust infrastructure absorb change without disruption. Those without it will be perpetually in reactive mode, treating each new regulation as a crisis rather than a standard they were already meeting.

Consider what this looks like in practice. A financial institution with mature governance in place—documented data lineage, pre-approved explainability frameworks, and automated bias monitoring—moves a new credit model from prototype to production in weeks, not months, while a competitor without those foundations waits in a compliance queue. That same infrastructure builds brand equity no marketing campaign can manufacture: As AI bias cases reach mainstream headlines, the ability to demonstrate audited, explainable systems becomes a durable differentiator. And as global regulatory requirements continue to accelerate, organizations with adaptable governance absorb each new mandate as a configuration update rather than a crisis. The organizations that will win the AI era are not those who moved fastest before governance caught up. They are the ones who built governance infrastructure that made moving fast safe in the first place.

Conclusion

This chapter opened with three questions that every organization deploying AI must be prepared to answer:

1. "What assumptions governed the outcome?" is answered by Pillar 1. Every model is trained on data that encodes assumptions about whose history to repeat, whose outcomes to optimize, and what success looks like. Data provenance makes those assumptions visible, auditable, and correctable before they cause harm. As the recruiting bias example illustrated, the cost of leaving assumptions invisible can be severe and swift.
2. "Can we prove its validity to a regulator, a board of directors, or an affected consumer?" is answered by Pillar 2. The stakeholder explainability framework defines precisely what proof looks like for each audience. The EU AI Act makes clear this is no longer a best practice; it is a legal obligation.
3. "Should the system have made that choice?" is answered by Pillar 3 and the ethical MLOps framework working together. Continuous guardrails define, in advance, the conditions under which a model is authorized to act and automatically route decisions to human review when those conditions are not met. Decision authority must be defined at the design table, before a model is built, by a team that includes risk, legal, and business leaders. The answer to "Should it?" is never left to the algorithm.

These three questions are not a threat to organizations that treat trust as infrastructure. They are an invitation to differentiate, to accelerate, and to lead. The AI era will not be won by the organizations with the most powerful models. It will be won by those who can look any stakeholder in the eye and say with full confidence, "We know what our systems do, we know why they do it, and we can prove it."

About the Author

Dr. Sheik M. Isaacs is an AI governance and enterprise data executive with 23-plus years of leadership in regulated financial services. Beyond the boardroom, he brings practitioner expertise into graduate classrooms at multiple universities, teaching MBA-level courses in AI, data governance, IT leadership, and organizational behavior. He also serves as a subject matter expert designing AI ethics and data science curricula at the university level. Dr. Isaacs holds a Doctor of Business Administration, five CIMP certifications across data management disciplines, and a Lean Six Sigma Black Belt. Drawing on decades of enterprise leadership and graduate-level instruction, he translates complex AI governance challenges into practical, actionable strategies for modern business leaders.

Email: sheik_isaacs@yahoo.com

LinkedIn: https://www.linkedin.com/in/sheik-m-isaacs

CHAPTER 14

AI CAN WRITE THE SCRIPT. IT CAN'T HANDLE THE AUDIBLE.

By Karlyn Kieffer
Sales Expert, L&D Strategist, Podcast Host
Springfield, Virginia

Luck is what happens when preparation meets opportunity.

—Seneca

When was the first time you woke up and thought, "Holy cow, I just did in twenty minutes what used to take me two hours"? And when did it first occur to you that all of that reclaimed time has not actually translated into the career leap you expected by now?

I work with sales professionals trying to accelerate their time to yes and increase their win rates. I work with leaders wondering why their teams are not making the gains they projected. And I work with corporate professionals across industries who are more productive than ever and still wondering why the promotion, the bonus, or the bigger opportunity has not shown up. Here is the hard truth nobody wants to hear: AI is an assistant, not a magic bullet. It can help you prepare. It cannot perform for you.

Let's take a quick inventory. How often, and for what, do you use AI? Drafting emails. Building presentations. Researching prospects. Screening candidates. Summarizing reports. Even finding a great restaurant in a new city that fits your dietary needs. AI usage is accelerating across nearly every corner of business and life, and for good reason. It takes us from under-prepared to ready to engage, faster than ever before. That is something to celebrate.

But here is what I have learned from years of coaching sales teams, leaders, and executives: without a sharp, mentally agile human who knows how to listen, read a room, and adjust in real time, AI becomes a crutch. And a crutch, leaned on long enough, quietly erodes the very competitive advantage it was supposed to build.

The Research Already Sees It Coming

This is not just a hunch. MIT Sloan's review of recent research on human-AI collaboration found something striking: There was no average "human-AI synergy" in the scenarios studied. In many cases, humans working alone, or AI working alone, outperformed the two working together. The pattern that mattered most was this: When humans performed better alone, adding AI helped. When AI performed better alone, adding humans hurt the result.[1]

Read that twice. The deciding factor was not the tool. It was the human's ability to bring something AI could not: judgment, context, adaptability. Human agility is what determines whether AI becomes leverage or noise.

Separately, the Harvard Gazette reported on MIT Media Lab research warning that heavy reliance on AI-driven solutions may contribute to cognitive atrophy, a gradual dulling of our critical thinking muscles.[2] And a review by IEEE Computer Society of related research found that frequent AI use correlated with weaker critical thinking skills overall, though higher education appeared to act as a protective buffer.[3] None of this means AI is the enemy. It means the muscle we stop using is the one we eventually lose.

For decades, what we casually call "soft skills," communication, presentation, objection handling, the ability to stay composed with a hostile or unpredictable audience, were treated as nice to have. They were the first line item cut from a training budget and the last one restored. AI is turning that on its head. Because here is what AI still cannot do: When you are in a sales pitch, a client conversation, a performance review, or a tense negotiation, and the person across from you says something you did not expect, AI is not there. It cannot read the tone shift in their voice. It cannot see the arms cross or the eyes drop. It cannot tell you what to say next. Only you can do that, in real time, while it is happening.

When the Demo Met Reality

A client invited me to sit in on one of his sales demo calls. The prospect had booked the meeting herself through a scheduling link, so there had been no formal discovery conversation beforehand. In its place, the rep did what more and more sales professionals now do: he researched her LinkedIn profile, ran a quick search on her background, and fed everything he found into AI. He asked it to help him anticipate what she would likely care about, what questions she might ask, and where he should focus the demo.

The output was thorough. It was also, on its face, a little confusing. Based on her profile, she did not look like an obvious fit for his product. He prepared anyway, built his demo around the best assumptions he and the AI could make together, and kept one thought in the back of his mind: "Expect a surprise."

The call began beautifully. He welcomed her, thanked her for the time, set the agenda, all delivered exactly as planned. He shared his screen, walked into the demo, and offered the assumptions he had carefully prepared about what she would want to see.

She stopped him. "Oh, I see how you'd think that," she said, "but you must have only found one of my online profiles. I'm calling about something else entirely."

In that moment, every assumption the AI had helped build, and every slide he had prepared around it, became irrelevant. This is where the real skill kicked in, and it had nothing to do with a tool. He stopped sharing his screen immediately. He gave her his full attention, went back to the beginning, and ran a real discovery conversation. What she described turned out to be exactly the kind of customer his product was built for. He restructured the entire demo on the spot, live, based on what he was hearing in real time. It was a home run.

AI did the legwork beautifully. It helped him walk into that call prepared instead of cold. But the moment the conversation went somewhere the research could not predict, the win came entirely from his ability to drop the plan, listen, and rebuild in real time. That is not a soft skill. That is the skill.

The Improv Exercise That Changes the Room

This is exactly why I build improvisation training into every program I deliver, not as an icebreaker, but as a core communication skill. Most people know "Yes, And," the classic improv exercise that teaches you to build on whatever you are given instead of evaluating, judging, and shutting down each idea as it comes out. It is useful, and I use it often.

But the exercise that consistently creates the fastest "aha" moment with skeptical business audiences is one called "Questions Only." The rule is simple: Participants can speak only in questions. No statements. No declarations. Just questions, back and forth.

In a sales context, this exercise is revealing almost instantly. Discovery, the ability to genuinely understand what a buyer needs, is the single biggest driver of moving someone from curious to interested to committed. Yet when sellers are forced to speak only in questions, they quickly discover how "me-centered" their normal conversations are. Many sellers ask questions designed to surface a specific keyword, the "magic word" that lets them launch into a pitch, rather than genuinely exploring what the buyer is dealing with.

After a few rounds of Questions Only, something shifts. Sellers ask second, third, and fourth-level questions they would never have thought to ask. They uncover what the buyer truly needs, not just what they initially said they wanted. They build trust faster because the buyer can feel the difference between being interviewed and being understood. Will anyone ever run an entire real conversation in questions only? No. But the exercise rewires a habit, from filling the air with your own words and agenda to genuinely asking, listening, and learning first.

The same exercise transforms leaders. Think about how many people you know who were excellent individual performers and got promoted into management without ever being taught how to manage people, how to give feedback that lands, how to motivate a team after a tough loss, or how to lead a mix of quiet, outspoken, and difficult personalities all at once. Questions Only will not fix all of that on its own, but it builds the foundation every great leader needs: curiosity. What makes this person tick? What is getting in their way? What is the best way to help them see it for themselves?

Our instinct as leaders is to jump straight to telling people what to do differently. But telling rarely creates lasting change. When someone arrives at their own answer through good questions, asking what worked, what they would do differently, what felt hard, they own that insight. And people rarely repeat a mistake they discovered and corrected themselves.

Who Wins as AI Reshapes the Workplace

I have worked with salespeople, executives, partner managers, and media spokespeople, and I do not see a consistent pattern of AI adoption by role, title, or generation. The pattern I do see among the people rising fastest has nothing to do with age, industry, or job title, and everything to do with approach.

The people winning with AI use it to do the legwork: finding credible, current sources, summarizing dense material into something they can act on, brainstorming options for a title, an angle, or a next step, and tightening an outline when their own thinking feels scattered. Then, they take that output and apply their own judgment, their own observations, and their own voice to produce something genuinely better than either the human or the AI could have produced alone.

I have also noticed that the strongest AI users add one small instruction to almost every prompt: "Ask me up to five questions that would help you produce a better result." AI is an extremely capable mirror, and like any mirror, it tends to reflect exactly what you hand it. Left unchecked, it can quietly reinforce your own blind spots and confirmation bias. Inviting it to push back, to ask what is missing, is itself a human skill, the same curiosity that powers great discovery conversations and great leadership.

If you're still not convinced that AI needs you, it's important to note that the very companies building the AI tools reshaping the workplace are still hiring sales reps and customer success managers, and they're doing so, aggressively. If anyone had the insight and incentive to replace those roles with AI, it would be them. Someone still has to handle the nuances.

So, who loses? Not a generation. Not a job title. The person who loses is the one who forgets that they are the ultimate value-add to AI, not the other way around. Your judgment, your emotional intelligence, and your ability to read a room are what turn AI from a generic tool into a genuine advantage. Lose sight of that, and AI does not replace you. It simply exposes how replaceable the parts of your work were that never needed you in the first place.

Two Hours, One Hundred Questions, and the Last Two

Years ago, long before AI could help anyone prepare for anything, I interviewed for a newly created role at the USTA. I was up against three internal candidates, each of whom had helped write the job description. I knew exactly how steep that hill was.

So, I prepared. For three days leading up to the interview, I wrote out one hundred questions I thought might come up and answered every one of them out loud, every day, until the answers felt effortless. The interview itself was a two-hour conversation with a five-person panel. Looking back, nearly every question they asked was one I had rehearsed.

Except the last two. One panelist asked how many other jobs I was interviewing for. Another asked my age, in Spanish, to test whether I had been honest about speaking conversational Spanish. I'm pretty sure it was the only question he remembered from his own high school Spanish class.

Neither question was on my list. After two hours of feeling like I was hitting every shot, I had to think on my feet at the buzzer. I told them this was the only opportunity I was pursuing. And in Spanish, I said, "Tengo 31 años, y sé que preguntaste eso para probar mi español": *I am 31 years old, and I know you asked that to test my Spanish.*

Preparation got me through 98 questions. The last two were handled by my continuous practice of dealing with the unexpected. Only the ability to stay calm and think clearly in uncontrolled conditions could do that. I got the job.

To be clear, preparation and improvisation are not competing strategies. They are partners. The deeper your preparation, the more confidently you can let go of the script when the moment calls for it. Why bring in a story that pre-dates AI? AI has made the preparation stage faster and more thorough than ever. That leaves more time for the improvisation stage so use it to help you own your conversations.

The Quiet Risk Worth Naming

Here is what concerns me most, not as a trainer, but simply as a person watching this unfold. Too many people are quietly losing the appetite for critical thinking. They write a prompt, get an output that feels close enough, and paste it into an email, a presentation, or a campaign. At most, they scan for an obvious AI tell. They rarely stop to ask: Is this accurate? Is this the right call? Is this what I think, or just what the machine produced?

Engaged listening, paying attention not just to what is said but to tone, pace, body language, and what is left unsaid, is becoming a rarer skill precisely because leaning on AI as a crutch is becoming more prevalent. The people who let that skill fade are not just at risk of a missed promotion. Over time, they risk becoming reactive rather than proactive, more comfortable following than leading, at the mercy of whatever direction momentum, or someone else's agenda, happens to be pointing.

That is not a dystopian prediction. It is a quiet, slow erosion, and it is entirely avoidable.

What to Do Differently After Reading This

You do not need to choose between AI and human skill. The professionals pulling ahead right now are doing both, deliberately. Here is the one shift to make in your very next high-stakes conversation: Let AI prepare you, and then do the human work required to own the conversation.

Use AI to research, summarize, brainstorm, and organize your thinking, exactly as it is designed to do. Then, before you walk into that meeting, that call, or that pitch, set the materials aside and ask yourself one question: "What happens if this goes completely off script?" Not to script an answer for every possibility, that is impossible, but to remind yourself that your job in the room is listen, adapt, and respond to the human in front of you.

Try one round of Questions Only before your next important conversation, even alone. Spend two minutes asking yourself only questions about the person you are about to meet with. What do they truly need? What might they be worried about? What have they not said yet that matters? You will walk in sharper, and more ready to hear them, than any prompt alone could make you.

AI is not going to slow down, nor should it. It is helping all of us move faster, research deeper, and prepare better than we ever could before. But speed and preparation have always been table stakes. What separates the professionals who thrive from the ones who stall out is what happens after the plan meets the moment.

The future belongs to the people who can leverage AI to help them prepare and adapt faster, pivot in the moment when it matters, and connect with the human in front of them in a way no machine ever will.

References

1. Michelle Vaccaro, Abdullah Almaatouq, and Thomas Malone, "When Combinations of Humans and AI Are Useful: A Systematic Review and Meta-analysis," *Nature Human Behaviour* 8 (2024): 2293–2303, https://doi.org/10.1038/s41562-024-02024-1; summarized in Brian Eastwood, "When Humans and AI Work Best Together—and When Each Is Better Alone," *MIT Sloan*, February 3, 2025, https://mitsloan.mit.edu/ideas-made-to-matter/when-humans-and-ai-work-best-together-and-when-each-better-alone.
2. Nataliya Kosmyna et al., "Your Brain on ChatGPT: Accumulation of Cognitive Debt When Using an AI Assistant for Essay Writing Task," arXiv preprint arXiv:2506.08872, June 2025, https://doi.org/10.48550/arXiv.2506.08872; reported in Liz Mineo, "Is AI Dulling Our Minds?," *Harvard Gazette*, November 13, 2025, https://news.harvard.edu/gazette/story/2025/11/is-ai-dulling-our-minds/.
3. Michael Gerlich, "AI Tools in Society: Impacts on Cognitive Offloading and the Future of Critical Thinking," *Societies* 15, no. 1 (2025): 6, https://doi.org/10.3390/soc15010006; summarized by IEEE Computer Society.

About the Author

Karlyn Kieffer is the founder and chief learning officer of Pro Tip Partners, a Washington, DC-based consultancy focusing on leadership and talent development. With more than twenty years of experience in sales enablement, learning and development, and instructional design, she has helped organizations, from startups to Fortune 100 companies, sharpen the human skills that drive business results. Before her career in corporate learning, she was a top-rated television news anchor in South Carolina and Georgia, experiences that shaped her lifelong belief in the power of staying composed and present under pressure. She is a Certified Professional in Training Management (CPTM) and a graduate of Georgetown University. Karlyn is the creator of the ACTION Sales Framework and host of the *Road to Greatness* podcast.

Email: Karlyn.kieffer@ProTipPartners.com

Website: www.ProTipPartners.com

LinkedIn: https://www.linkedin.com/in/karlyn-kieffer-cptm/

CHAPTER 15

FROM DATA TO INTELLIGENCE: SCALING ENTERPRISE AI INTO MEASURABLE BUSINESS TRANSFORMATION

By Vivek Kompella
Data & AI Leader, Strategist
San Jose, California

It is not the strongest of the species that survives, nor the most intelligent. It is the one most adaptable to change.

—Charles Darwin

We live in a digital world with an abundance of data. Every click, every transactional log, every sensor ping, and every customer interaction leaves behind a massive, permanent digital footprint. For years, boardrooms have been told that data is their most valuable asset, the primary currency of the modern economy. Naturally, executives want to monetize this resource to accurately measure operational performance, deeply analyze customer behavior, and leverage those behavioral signals to aggressively drive revenue growth. Yet, as they look across their vast data lakes and cloud repositories, a frustrating paradox emerges: Most organizations are drowning in data but starving for intelligence. They are data-rich but insight-poor.

This dynamic has intensified as the artificial intelligence marketplace undergoes a massive structural shift. AI is no longer a siloed, highly technical discipline restricted to data scientists; it has rapidly matured into a mainstream skill accessible to a wider range of job roles. Recognizing this widespread accessibility, forward-thinking leaders are eager to capitalize on the momentum and finally realize the true value of their data assets. There is a clear understanding across the C-suite that modern machine intelligence provides the unprecedented processing speed required to turn passive, unstructured information into active market advantages.

However, executing this strategy requires overcoming significant operational bottlenecks where data and AI intersect. Organizations routinely struggle with strategic prioritization, separating high-yield use cases from temporary technology hype. Internal alignment presents an ongoing friction point, as companies find it incredibly difficult to get their data governance policies, IT architecture, and specific business units to move in lockstep. Furthermore, the spiraling financial realities of AI infrastructure spend mean that many organizations find themselves burning through budgets on localized, isolated pilot projects that never successfully transition to true enterprise scale.

Moving past these operational hurdles requires a deliberate, structured framework. To bridge the gap and capture the full return on your technology investments, the enterprise must simultaneously scale its data infrastructure alongside its organizational culture. This chapter examines that exact operational journey, providing a practical roadmap to move past basic data collection and achieve the genuine enterprise intelligence that drives measurable business transformation.

The Intelligence Gap: Why Most AI Investments Underperform

Before we can discuss how to scale AI into genuine business transformation, we need to be honest about why so many efforts stall. The failure is rarely technical. The models work. The APIs are available. The cloud infrastructure scales. The failure is almost always organizational: a mismatch between what AI can do and how the enterprise is structured to receive it.

Data as Infrastructure

Every organization has data. Very few have trusted data that is ready to be useful. For decades, enterprise data has been treated as a byproduct of operations rather than a strategic asset, and it's been scattered across CRMs, ERPs, and a patchwork of legacy systems, with integration expensive and governance inconsistent.

AI changes the economics of data significantly, but it does not change the fundamental requirement for data quality. A language model trained on inconsistent, poorly labeled enterprise data will produce confusing, confidently incorrect outputs. The principle holds: Intelligence is only as good as the data it reasons from.

The enterprises winning with AI made the unglamorous investment first. They built a semantic layer—a unified, governed representation of what the business knows, expressed in terms that are consistent and machine-readable—and established clear ownership for the quality of each data domain. This work is not exciting and rarely makes a press release. But it is the difference between AI that produces noise and AI that produces signal.

The Measurement Trap

There is a second reason AI investments underperform, and it is subtler: Most organizations measure AI activity rather than AI value, the number of models deployed, not the business outcome those models actually moved.

This discipline cannot be a post-implementation fix. Before deploying any AI capability at scale, the organization should be able to answer three questions: What decision does this AI improve? What is the baseline today without it? And what measurable gap is it expected to close?

Every AI use case should map to one of four value levers: cost savings, customer experience, innovation, or productivity. Naming the lever up front forces clarity on which metric will prove the investment worked: new capabilities for innovation, improving customer experience, increase in productivity, and margin for cost savings.

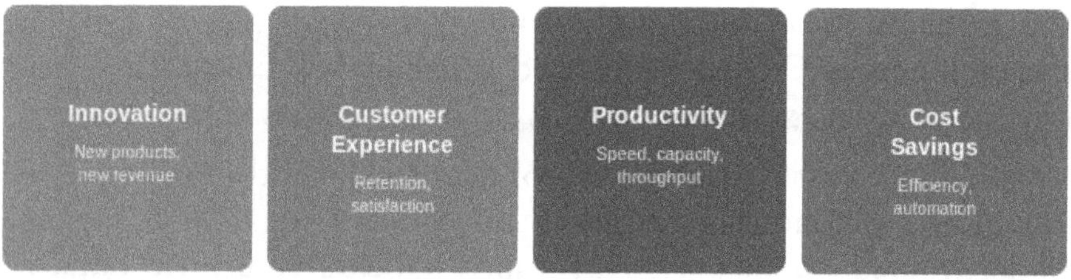

Figure 1: The four value levers of measurable AI transformation.

The Takeaway: If you cannot define the outcome before you deploy the model, you will not be able to attribute the outcome after. Measurement is not an afterthought. It is the architecture of accountability.

Solving the Wrong Problem, Brilliantly

The third and most common reason AI investments underperform is that teams pick use cases for their technical novelty rather than their business priority. In this way, they build something genuinely impressive that solves a problem no executive was asking about, and so what's produced never earns the sponsorship or budget to scale past the pilot.

The Strategic Imperative: Why Waiting Is Not an Option

Some leaders, cautious by temperament or scarred by past technology investments that overpromised, are inclined to wait—let others experiment, let the ROI cases get clearer. This instinct is understandable. It is also increasingly dangerous.

The Compounding Advantage

AI creates compounding advantages in a way most technologies do not. A better software platform can be observed and replicated by competitors. Superior AI capability is harder to reverse-engineer because the capability is not just the model, but the data flywheel feeding it, the organizational learning embedded in how teams use the outputs, and the institutional knowledge built into the workflows around it.

Amazon's recommendation engine is not just an algorithm. It is 25 years of customer behavior data, organizational learning about how to present products, and continuous experimentation infrastructure that improves the system millions of times per year. A competitor building a similar algorithm today starts from zero. The gap is not technical. It is temporal.

This dynamic is now playing out across financial services, manufacturing, healthcare, and logistics alike: Early movers are accumulating data assets and AI-native processes that are increasingly difficult for slower organizations to replicate. The window closes a little more each quarter.

The Architecture of Transformation: Three Pillars That Scale

What separates enterprises that achieve genuine AI transformation from those that accumulate an expensive portfolio of stalled pilots? It is not the sophistication of their models or the scale of their cloud contracts. It is the presence of three interconnected organizational pillars that, taken together, create the conditions for AI to move from experiment to enterprise capability. Each pillar is necessary. None is sufficient on its own. And the order in which they are built matters as much as the pillars themselves.

Figure 2: The three pillars of enterprise AI transformation.

Pillar 1: Trusted Data Foundation

Every enterprise AI ambition rests on one unglamorous prerequisite: data that can actually be trusted. Not data that exists—virtually every organization has that in abundance—but data that is accurate, consistent, well-governed, and meaningfully connected across the systems that generate it. Without this foundation, AI systems produce outputs that domain experts rightly dismiss, executives cannot act on, and auditors flag as unreliable. The technology is only as intelligent as the information it reasons from.

Think about what it takes to build an AI system that predicts customer usage and hands the sales team a ranked list of leads. It is genuinely powerful when it works: faster than any analyst, more consistent than any intuition, operating at a scale no team could replicate manually.

But here is the question I always ask before a single line of model code gets written: Do we trust the data this will run on? In one engagement, the honest answer was no. Usage data was captured differently across product lines. Customer records lived in systems that had never been reconciled. Definitions of what counted as an active account varied by region. The AI would have produced a beautiful, confident, completely unreliable output, and the sales team would have known it within a week.

We spent three months on the data before we touched the model. The forecast we eventually delivered was not the most sophisticated model I have ever seen. It was the most trusted one, and that made all the difference.

Building a trusted data foundation means investing in four things simultaneously. First, data quality: establishing the standards, monitoring, and remediation processes that ensure data is fit for the decisions that depend on it. Second, data integration: connecting the disparate systems, lakes, and warehouses that each hold fragments of what the organization knows into a coherent, accessible whole. Third, a semantic layer: a shared business glossary that defines what every metric, entity, and event means consistently across functions, so that revenue in finance means the same thing as revenue in sales and operations. Fourth, data lineage and traceability: the ability to follow any AI-generated output back to the data that produced it, which is essential not just for debugging but for the explainability requirements that responsible AI demands.

This investment is not exciting. It does not make headlines. But it is the single greatest predictor of whether an organization's

AI initiatives will deliver lasting value or fade after their initial demonstration. The enterprises that have built this foundation find that AI deployments move faster, cost less to maintain, and earn the trust of the people whose decisions they inform. Those that skip it find themselves rebuilding under pressure expensively and with diminishing organizational credibility.

The Takeaway: You cannot build intelligent systems on unreliable information. The data foundation is not the starting point of AI transformation. It is the entire ground beneath it.

Pillar 2: Center of Excellence

If the data foundation is the ground, the center of excellence (CoE) is the structural frame built upon it: the organizational engine that turns isolated AI ambition into coordinated, scalable enterprise capability. The CoE is not a single team in a corner of the IT department. It is a cross-functional governing body with a mandate that spans three interconnected domains: building the shared infrastructure, setting the enterprise standards, and owning the accountability for how AI is developed and deployed responsibly across the organization.

> *The hardest governance problems are rarely the ones where someone is doing something obviously wrong. They are the ones where someone is doing something that works, just not safely. That was the situation when I discovered a senior sales leader had rolled out a free-tier AI tool across his team. His team loved it. They were using it every day for research, email drafting, and call prep. And they were feeding it sensitive customer data, pipeline context, and competitive intelligence with no visibility into where any of it was going.*
>
> *When the CoE team raised the security issue directly, the leader was candid: Internal AI was not integrated into the CRM where his team worked, so no one used it. Fix that, he said, and he would make the switch.*

So CoE spent time with the reps to understand exactly what they valued about the external tool, then worked with data and platform teams to build the equivalent in-house: a company research agent, integrated directly into the CRM, shaped by feedback from the team who would use it. Four weeks. One working product. A firm agreement with the leader on timelines and adoption milestones.

The shadow AI disappeared not because we mandated it out of existence, but because we replaced it with something better. That is what a CoE actually does.

On the infrastructure side, the CoE centralizes what individual business units should never have to build independently: AI platforms, reusable model components, centralized compute, and the integration layer connecting AI to enterprise data and workflows. This is not control for its own sake. It eliminates the duplication and technical debt that accumulates when every team builds its own AI stack from scratch.

On the standards and governance side, the CoE owns the enterprise's responsible AI mandate. It's a deliberate structural choice. By embedding this mandate inside the CoE, the same team accountable for shipping AI capability is also accountable for shipping it responsibly: fairness and bias auditing, explainability, privacy and security standards, and the human oversight mechanisms that let AI systems be challenged, corrected, and retired when they diverge from their intended purpose.

The CoE also drives business alignment, bridging the gap between technology teams who know what AI can do and business units who know what problems are worth solving. Acting as an internal consulting function, it helps domain experts translate their knowledge into well-formed AI use cases, preventing investment in technically impressive AI that solves the wrong problems.

The Takeaway: The CoE is where strategy becomes capability, where principles become practice, and where AI ambition finds the organizational home to survive contact with reality.

Pillar 3: Organization and Culture

Technology adoption at enterprise scale has always been a human problem disguised as a technical one. AI makes this dynamic more acute, not less. The organizations furthest ahead in AI transformation are not those with the most sophisticated models. They are those that have done the hard, patient work of building a culture where humans and AI systems work together effectively, where experimentation is encouraged, where failure is treated as data rather than defeat, and where the benefits of AI are distributed across the workforce rather than concentrated in a small technical elite.

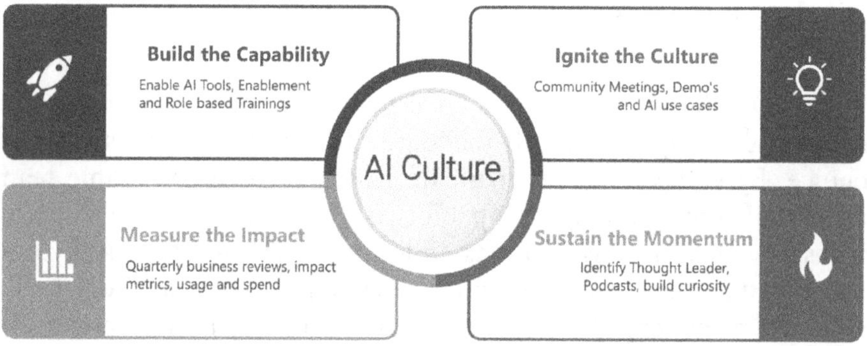

Figure 3: The AI Adoption Flywheel: Build, Ignite, Sustain, Measure.

Changing how an entire organization thinks about and works with AI is not a program you launch and complete. It is a system you build and keep running. The AI Adoption Flywheel is a practical framework for doing exactly that: four interconnected motions that, when executed in sequence, create self-reinforcing momentum. Each motion feeds the next. Over time, the organization does not need to be pushed toward AI adoption. It pulls itself.

1. Build the Capability

The starting point is ensuring the organization has the knowledge to engage with AI meaningfully. This begins with broad awareness: a literacy program helping every employee understand what AI is, what it can and cannot do, and why it matters. But awareness alone is not capability. The second layer is role-based enablement: targeted training that teaches each function—sales, finance, operations, marketing—to use AI in the context of their actual daily work, not generic tutorials.

2. Ignite the Culture

Capability without community stalls. A monthly internal community meeting—where teams share real use cases in a show-and-tell format—shows people what is actually possible inside their own organization, by people they know and trust. These sessions also surface the legacy processes that AI could modernize but that no one has yet questioned. When a colleague shows how AI cut their reporting time from four hours to twenty minutes, the person next to them starts rethinking their own process. That is how culture shifts, not through mandate, but through visible, relatable proof.

3. Sustain the Momentum

Every organization has people who are naturally curious and naturally influential among their peers. The third motion is about finding them and giving them a platform: internal AI thought leaders, podcasts hosted by practitioners for practitioners, hackathons that give teams protected time to experiment on real business problems. These are not expensive programs. They are permission structures, signals that curiosity is valued and the people closest to the work are trusted to shape how AI enters it.

4. Measure the Impact

The flywheel only keeps turning if the organization can see that it is working. Quarterly stakeholder business reviews—bringing together business leaders, the CoE, and AI platform teams—create the accountability rhythm that keeps AI transformation on track. Three questions anchor every review: Where is AI being used and at what scale? What is the organization spending, and is it being spent wisely? And most importantly: What measurable business value has AI delivered this quarter? The answers to those questions determine where to invest next, what to scale, and what to retire. They also generate the evidence that justifies continued investment to the board, moving AI from a line item that requires faith to a capability that speaks for itself in the numbers.

The Takeaway: Culture does not change because leadership announces it. It changes because individuals see proof, find community, feel empowered, and measure progress. The flywheel builds all four, and once it is turning, it does not need to be pushed.

The Path Forward: From Experiment to Civilization

We are not in the early stages of an AI experiment. We are in the early stages of a fundamental restructuring of how enterprise value is created. The organizations that recognize this and respond with the seriousness, discipline, and human wisdom the moment demands will not merely survive the disruption. They will define what comes next.

The path is not mysterious. Build the data foundation before the AI systems. Structure your teams around outcomes, not functions. Choose use cases with discipline and measure results with rigor. Above all, resist treating AI as a technology problem; it never was one. It is a strategy problem, a culture problem, and ultimately a leadership problem: the challenge of moving an organization from what it has always known to what is now possible.

The disruption is real and the window is open right now. The question is not whether AI will transform your industry. It is whether your organization will be the one doing the transforming or the one being transformed by someone else.

The data is there. The technology is ready. The only question left is whether your organization has the leadership to close the gap between what it knows and what it can become.

About the Author

Vivek Kompella is a strategic and visionary data and AI leader with nearly two decades of experience driving enterprise data and AI strategy, including building and leading AI centers of excellence. He has a proven track record of architecting data platforms, deploying AI-powered solutions, and guiding geo-distributed teams to scalable, cost-effective, measurable business results. Connect with Vivek.

Email: kompella_anand@yahoo.com

LinkedIn: https://www.linkedin.com/in/vivekkompella/

CHAPTER 16

THE HUMAN STILL HAS THE HARDEST JOB

By Byron M. Lowens, PhD
HCI Researcher and Human-Centered AI Educator
Monroe, Louisiana

The first principle is that you must not fool yourself,
and you are the easiest person to fool.

—Richard P. Feynman

It is a Tuesday morning, and Maya, a clinic administrator, has a decision to make before her leadership meeting. Her team is short-staffed, missed appointments are creating pressure on the schedule, and a vendor is offering an AI tool that predicts which patients are most likely to miss their next visit. The pitch sounds simple: Flag the high-risk patients so the clinic can act early. So, Maya opens a chatbot and asks whether her clinic should adopt the tool. Within seconds, she gets a clean, organized answer with benefits, risks, and a recommendation laid out more neatly than most people could produce under that kind of pressure.

The answer is useful, but this is where the real work begins. A list of patients who are "likely to miss appointments" is rarely just a list of careless people. More often, it is a list of people dealing with the hardest barriers: unreliable transportation, shifting work schedules, childcare that falls through, or past experiences that have made the clinic feel unwelcoming. Depending on how the clinic uses that list, the same prediction could lead to a helpful reminder call and a ride voucher, or it could quietly push the most vulnerable patients further away from care. The AI can produce the prediction in seconds, but it cannot decide what the prediction means, who it might harm, or whether using it is fair. The AI also cannot carry the responsibility if the clinic gets that decision wrong.

I have spent my career as an HCI researcher and educator studying moments exactly like this one. A capable person meets a capable tool and realizes that the hardest part was never just using the tool. Whether through teaching students, advising teams, or helping organizations think through the social impact of technological systems, I return to the same principle: Technology decisions are human decisions first. AI has not changed that. It has raised the stakes, because it can make unfinished thinking look complete. A fluent answer can feel complete. A confident tone can feel reliable. But fluency is not the same as judgment. An AI tool can organize information, surface risks, and suggest options, but it does not know Maya's clinic, her community, or the moral weight of the decision in the way a responsible human decision-maker must. The fluency is real. The accountability is not.

That is the central tension of the AI moment. AI can reduce effort, but it can also reduce reflection. The same tool that saves Maya twenty minutes can tempt her to skip the twenty minutes of thinking the decision requires. The faster and smoother the answer, the more disciplined the human must become. That is not a complaint about the technology. It is a description of the new role it creates for us.

AI Is Not Just Replacing Tasks. It Is Reshaping Judgment

Most conversations about AI and work begin with the fear of replacement: which jobs will disappear, which tasks will be automated, and who will be left behind. Those questions matter, but if we only focus on the labor market, we miss the silent disruption happening inside human judgment. AI is not only changing what we can do. It is changing how we search for information, how we write, how we weigh evidence, how we learn, and how we decide whom and what to trust. Realistically, the risk is not only job displacement, it is judgment displacement. Judgment displacement is the slow, often invisible process of handing the hard parts of thinking to a system that performs them faster but does not understand them. When AI drafts the email, frames the analysis, and proposes the recommendation, the human role can quietly shrink from author to approver. Approving feels like deciding, but usually it is not. This is the quiet failure that keeping a human "in the loop" is supposed to prevent, and, as we will see, the reason it so often does not.

Human factors researchers saw this pattern long before generative AI arrived. In a now-classic argument called the ironies of automation, Lisanne Bainbridge (1983) showed that as systems become more automated, the people overseeing them can become less practiced, less attentive, and less prepared to intervene at exactly the moment something goes wrong.[1] Automation does not simply remove effort. It relocates it, often to the worst possible moment, when a human who has stopped paying close attention is suddenly expected to step in and exercise expert judgment. Generative AI extends this irony from cockpits and control rooms into everyday knowledge work: into the email, the memo, the lesson plan, and the hiring screen.

This is why I find human-centered AI (HCAI) more useful than the familiar debate about replacement. HCAI is not simply about making AI easier to use. It is about making sure AI systems are designed, evaluated, and adopted in ways that preserve human agency, protect people from harm, and keep accountability visible.[2] When I teach this in my courses, I remind students that the goal is not simply to make AI more powerful. The goal is to make sure powerful systems still serve human needs, human values, and human responsibility.

The remainder of this chapter is about how to do that in practice. It offers a more disciplined way to work with AI: Frame the real problem before asking for answers, collaborate with the system instead of deferring to it, interrogate what it produces, decide with human judgment and ethics, and remain accountable for the result. AI is reshaping how we think. Whether that reshaping strengthens or weakens our judgment depends on what we continue to ask of ourselves while we use it.

The Five Responsibilities AI Cannot Own

The faster and smoother the answer, the more disciplined the human has to become. That discipline is easy to reduce to a vague instruction to "be careful," but it is much more specific than that. AI can help produce answers, but there are five harder responsibilities it cannot own for us. These are the responsibilities that decide whether an answer is useful, fair, and worth acting on.

Responsibility 1: Framing the Problem: Before Maya opens the chatbot, someone must decide what question is really being asked. Is the clinic trying to reduce missed appointments, or is it trying to help patients who are struggling to reach care? Those sound similar, but they can lead to very different systems. AI will answer the question it is given. It may even suggest a better framing, but it cannot be responsible for deciding what problem the organization should be solving. A precise answer to the wrong question is one of the most expensive things an organization can produce.

Responsibility 2: Understanding the Context: A model learns patterns from large amounts of text and data, but those patterns do not automatically carry the lived context of a specific place. It does not know that the bus route serving Maya's patients fails before dawn or that one family stopped coming after a bad experience. That knowledge lives with the people who work in the clinic and the people it serves. Maya's job is to bring that context into the decision and to notice when a confident recommendation is built without it.

Responsibility 3: Interpreting Consequences: A prediction that a patient may miss an appointment is just a number until a human decides what happens next. The same flag can trigger a supportive reminder and a ride voucher, or it can quietly move the most vulnerable patients down the priority list. AI can estimate likelihoods, but it cannot decide what those likelihoods should mean once they touch people's lives.

Responsibility 4: Choosing Values: Deciding whether to optimize for clinic efficiency or patient equity is not just a calculation; it is a judgment about what matters and who counts. AI can lay the tradeoff in front of Maya in a clean table. It cannot make the choice, because the choice expresses a commitment, and commitments belong to people.

Responsibility 5: Accepting Responsibility: Once the decision is made, someone has to own it, defend it, and answer for it if it causes harm. This is the responsibility that cannot be delegated to a tool. A system can be wrong, but it cannot be accountable.

When I teach HCAI, I remind students that the first challenge is rarely whether people can learn the tools. Most people can. The harder challenge is whether they can do these five things well, and whether they recognize that the smooth answer on the screen has done none of them. Well-designed systems can help by making their limits visible and allowing people to correct, challenge, and override them.[3] But design only creates the opening. Framing, context, consequences, values, and responsibility remain ours. This is the part the hype skips. AI can expand what a person can do, but it cannot relieve that person of the responsibility to do it well. Once we see that the human job is this specific, a familiar question starts to look incomplete. The question is not simply whether we can trust AI. It is what we should trust it to do, under what conditions, and with what level of human judgment still in place.

Trust Is Not the Goal. Calibration Is the Goal

At this point, many people ask what sounds like the responsible question: "Can I trust AI?" But that question is a bit broad. Trust is not a yes-or-no decision about a tool. It is a judgment about a specific task, in a specific situation, with specific stakes. The better question is not whether Maya should trust AI in general. The better question is what she should trust this system to do, based on what evidence she has, and with how much human review.

Researchers have studied this problem for decades, and they have names for the ways it goes wrong. Automation bias is the tendency to accept an automated recommendation too quickly, especially when the system sounds confident, technical, or authoritative.[4] Overreliance happens when people depend on a system beyond what it can support, treating a helpful estimate as if it were a settled fact. But the opposite problem matters too. Under reliance happens when people reject useful support because they are afraid of the tool, confused by it, or unsure when it is appropriate to use. Both are failures of calibration. One gives the system too much authority, the other refuses help when help would be useful. Neither is the goal. This is something I see often from my students and clients. Some people are so impressed by what AI produces that they begin to treat its answers as authority, while others are so skeptical or intimidated that they avoid it even when it could genuinely help. HCAI rejects both extremes. The goal is not blind trust, and it is not blanket refusal. The goal is what John Lee and Katrina See (2004) called appropriate reliance: matching the level of trust to the task, the evidence, the stakes, and the known limits of the system.[5]

That changes what Maya is actually deciding. The problem was never whether to trust AI. The problem is trusting it at the wrong time, for the wrong task, at the wrong level, and without enough human review. For a low-stakes draft, fast reliance may be fine because an error is cheap and easy to catch. For a prediction that could sort vulnerable patients, that same speed becomes a liability. In that case, the output should be treated as one input into a human decision, not the decision itself.

This is not a hypothetical worry. In a widely cited study from 2019, Ziad Obermeyer and colleagues examined a commercial algorithm used to flag patients for extra care and found that it systematically underestimated the needs of Black patients.[6] The reason was subtle. The system had been trained to predict healthcare costs as a stand-in for healthcare needs, and because less money had historically been spent on Black patients with the same level of illness, the model learned to treat them as healthier than they were. The algorithm was not malicious. It was faithfully reproducing a pattern in the data, and that pattern carried the inequities of the system that produced it. This is what researchers mean by algorithmic bias: A model can be statistically accurate and still be unfair, because the data it learns from is a record of a world that was never neutral.[7] A fluent, confident recommendation can hide this completely. Fairness is not something a system certifies on its own. It is a judgment a person has to make about who a prediction helps and who it quietly harms.

The same logic applies beyond healthcare. In hiring, an AI tool that summarizes applications can help organize information, but it becomes dangerous the moment its ranking is treated as a measure of human potential. A resume is not the person, and a model's score is not a measure of worth. The more consequential the decision, the more the system's output must be questioned, placed in context, and kept subordinate to human responsibility.

This is where keeping a human in the loop actually matters. Human-in-the-loop design draws from a long tradition in human factors of keeping people in supervisory control of automated systems.[8] But the phrase can become empty if we are not careful. A person who simply approves every recommendation is not meaningfully in the loop. They are just giving the system a human signature. The human role needs to be active: reviewing outputs, catching ambiguous cases, questioning weak recommendations, and holding the authority to intervene before a recommendation becomes an action.

Trust, then, is not a feeling we either give or withhold. It is something we calibrate case by case. The goal is to keep people meaningfully involved in the decisions that carry the most weight, especially when those decisions affect someone's access, opportunity, care, or dignity.

The Human-Centered AI Compass

The responsibilities we have traced throughout this chapter explain what stays human. The Compass turns those responsibilities into something you can use in the moment, with human judgment at the center and five bearings to check before you act: Frame, Collaborate, Interrogate, Decide, and Account.

Bearing 1: Frame: Before you ask AI for an answer, decide what problem you are actually solving. Maya's choice between reducing missed appointments and helping patients reach care is made here, before a single prompt, and it shapes everything downstream.

Bearing 2: Collaborate: Use AI to expand your thinking, not to end it. Ask it to surface options, draft, organize, and challenge you, while keeping the system as one voice in the room rather than the voice that decides. This is calibrated reliance in practice. It is where a human stays meaningfully in the loop instead of simply signing off.

Bearing 3: Interrogate: Treat a fluent answer as a starting point, not a verdict. Ask what is missing, whose perspective is absent, what the system could not know, and where bias might be hiding. This is also where explainability matters. Many systems can now offer some account of how they reached an output, and the field of explainable AI has worked to make model reasoning more transparent and open to inspection.[9] A good explanation can help you catch a flawed assumption and judge how much weight a recommendation deserves, but an explanation is not the same as a justification. A system can give a clear, confident rationale for a conclusion that is still wrong, and a polished explanation can increase your trust in a bad answer rather than expose it. Transparency is a tool for interrogation, not a replacement for it. The smoother the output, the harder you press.

Bearing 4: Decide: Bring what the machine cannot: domain knowledge, ethical judgment, awareness of who is affected, and a sense of what the situation truly demands. The recommendation is an input. The decision is yours.

Bearing 5: Account: Own the outcome. Be ready to explain why you decided as you did and to answer for the consequences, because responsibility is the one thing that never transfers to the tool.

None of this requires becoming an engineer. It requires becoming a more deliberate human in an AI-shaped world, which is the real work and the real opportunity. The people who thrive in the AI transformation will not be those who use AI the most. They will be those who learn to think with it while keeping judgment, ethics, and accountability where they belong.

The AI transformation will reward people who can think with machines without letting machines think for them. Human-centered AI is not a soft concern. It is the discipline of making sure that as AI grows more powerful, the humans using it remain thoughtful, responsible, and accountable for the decisions that shape people's lives.

References

1. Lisanne Bainbridge, "Ironies of Automation," *Automatica* 19, no. 6 (1983): 775–79, https://doi.org/10.1016/0005-1098(83)90046-8.
2. Ben Shneiderman, "Human-Centered Artificial Intelligence: Reliable, Safe & Trustworthy," *International Journal of Human–Computer Interaction* 36, no. 6 (2020): 495–504, https://doi.org/10.1080/10447318.2020.1741118.
3. Saleema Amershi et al., "Guidelines for Human-AI Interaction," in *Proceedings of the 2019 CHI Conference on Human Factors in Computing Systems* (New York: Association for Computing Machinery, 2019), 1–13, https://doi.org/10.1145/3290605.3300233.
4. Raja Parasuraman and Dietrich H. Manzey, "Complacency and Bias in Human Use of Automation: An Attentional

Integration," *Human Factors* 52, no. 3 (2010): 381–410, https://doi.org/10.1177/0018720810376055.

5. John D. Lee and Katrina A. See, "Trust in Automation: Designing for Appropriate Reliance," *Human Factors* 46, no. 1 (2004): 50–80, https://doi.org/10.1518/hfes.46.1.50_30392.
6. Ziad Obermeyer, Brian Powers, Christine Vogeli, and Sendhil Mullainathan, "Dissecting Racial Bias in an Algorithm Used to Manage the Health of Populations," *Science* 366, no. 6464 (2019): 447–53, https://doi.org/10.1126/science.aax2342.
7. Ruha Benjamin, *Race after Technology: Abolitionist Tools for the New Jim Code* (Medford, MA: Polity Press, 2019).
8. Thomas B. Sheridan, *Telerobotics, Automation, and Human Supervisory Control* (Cambridge, MA: MIT Press, 1992).
9. Finale Doshi-Velez and Been Kim, "Towards a Rigorous Science of Interpretable Machine Learning," arXiv preprint arXiv:1702.08608, 2017, https://arxiv.org/abs/1702.08608.

About the Author

Byron M. Lowens was born and raised in northeastern Louisiana. He earned his bachelor's degree in electronics and electrical engineering from Southern University and A&M College, then a master's in computer science from Virginia State University, before completing his PhD in human-centered computing at Clemson University. His doctoral research focused on interaction techniques for in-the-moment privacy control of wearable data, asking how people decide what their devices should share and when.

After his doctorate, Byron served as a postdoctoral research fellow at the University of Michigan, where he led large-scale, mixed-methods studies of privacy behavior, the risks of wearable health data, and the gaps between what people say about privacy and how they actually behave on social media platforms. Today he is a lecturer in the Luddy School of Informatics at Indiana University Indianapolis, where he teaches graduate and undergraduate courses in human-computer interaction, human-centered AI, collaborative human-AI systems, ethical technology design, the psychology of HCI, social computing, and ubiquitous computing. He is also the founder of Genius User Experience LLC, where he provides user research, usability testing, design strategy, and AI consulting focused on responsible, human-centered AI adoption.

Across all of it, Byron's work examines how emerging technologies shape human judgment, privacy, trust, equity, health, and accountability.

Email: bmlowens@gmail.com

LinkedIn: https://www.linkedin.com/in/bmlowens/

CHAPTER 17

AI AND THE NEXT BREAKTHROUGH ECONOMY: EMPOWERING ENTREPRENEURS AND ENTERPRISES

By Mark L. Madrid
Economic Architect for America's Next 250
Austin, Texas

The future is not something we enter.
The future is something we create.
—Leonard I. Sweet

Every generation inherits a challenge. Ours inherits a breakthrough.

The steam engine expanded human muscle. Electricity expanded human productivity. The internet expanded human connectivity. Artificial intelligence is different. AI expands human capability itself. The question is not whether AI will reshape the economy. The question is who will help shape AI.

I have spent three decades helping build bridges between entrepreneurs, institutions, communities, and opportunities. From Wall Street to Washington, DC, from America's rural communities to global stages, I have witnessed firsthand how breakthrough moments reshape economies and redefine what is possible. Today, we are standing at the beginning of the most significant economic transformation of our lifetime.

As a former associate administrator of the US Small Business Administration, I traveled extensively across America meeting entrepreneurs, manufacturers, farmers, veterans, innovators, educators, and community leaders. I learned a lesson that has stayed with me throughout my career. The greatest challenge facing most organizations is not a lack of talent. The greatest challenge is a lack of leverage and lack of access to expertise, knowledge, networks, and economic opportunity. Artificial intelligence is changing all four.

At my company Breakthrough Mavens, our mission is simple: To energize businesses, enterprises, and leaders to transform potential into breakthrough execution. Our vision is equally ambitious: To energize America's Next 250 by powering a future-ready America where breakthrough entrepreneurs flourish, enterprises lead, and communities prosper.

That vision is not ultimately about technology. It is about people. It is about prosperity. It is about possibility. And it is about ensuring that the benefits of the next economic era are broadly shared.

This AI breakthrough belongs to everyone.

The Economic Opportunity Before Us

Economists and business leaders increasingly agree that AI represents one of the largest economic opportunities in modern history. AI could contribute $15.7 trillion to the global economy by 2030.[1] Generative

AI alone may create $2.6 trillion to $4.4 trillion in annual economic value.[2]

Looking further ahead, economists anticipate that the global economy could exceed $200 trillion by 2050,[3] fueled in part by advances in artificial intelligence, automation, robotics, quantum computing, and new forms of human-machine collaboration. Yet the greatest AI opportunity in America may not be found in Silicon Valley. It may be found among America's more than 36 million small businesses.[4] The next breakthrough economy will not be built by a privileged few. It will be built by millions.

This AI breakthrough belongs to everyone.

Those numbers are impressive. But the real story is not the technology. The real story is what technology makes possible. The real story is about the entrepreneur in a rural community gaining access to capabilities once reserved for multinational corporations.

The real story is a nonprofit expanding its impact without dramatically increasing overhead. The real story is a manufacturer becoming more competitive, a teacher becoming more effective, a healthcare provider becoming more efficient, and a small business owner becoming more profitable. AI is not replacing human potential. It reveals how much potential has been sitting idle.

This AI breakthrough belongs to everyone.

The Breakthrough Economy

Throughout my career, I have been fascinated by the relationship between innovation and prosperity. Economies grow when ideas become action. Communities prosper when enterprises create opportunity. Nations advance when entrepreneurs solve meaningful problems.

We are entering what I call the Breakthrough Economy. The Breakthrough Economy is not defined solely by technology. It is defined by amplified human potential.

In this new era, competitive advantage will belong to organizations that combine human ingenuity with intelligent systems. Success will not belong simply to those with the most technology. Success will belong to those who best integrate technology with integrity, purpose, leadership, culture, trust, and execution. Technology creates acceleration. Leadership creates direction.

Intelligence is becoming abundant. Wisdom remains scarce. The winners of the AI economy will not simply be the smartest organizations but rather the most adaptable.

Energy. Discipline. Grit.

Technology alone never guarantees transformation. Humans will drive outcomes. That is why three principles remain central to my core to my business philosophy: Energy. Discipline. Grit.

Energy fuels action. It creates momentum. It inspires innovation and encourages organizations to move beyond fear and into possibility. Discipline transforms intention into execution. It turns strategy into outcomes. Grit sustains momentum when uncertainty inevitably appears.

AI can accelerate productivity, but it cannot replace purpose. AI can generate recommendations, but it cannot replace leadership. AI can create possibilities, but it cannot replace integrity and courage. Technology amplifies. Integrity and character determine direction.

The Honeycomb Economy

As I have reflected on the future of economic development, one image continues to come to mind: the honeycomb. Honeybees have demonstrated one of nature's most extraordinary systems of value creation. No single bee builds the hive alone. No bee possesses all the answers. Yet together they create something remarkable.

The future economy will increasingly resemble a honeycomb of entrepreneurs, enterprises, community colleges, universities, governments, communities, AI agents, and human intelligence. All connected; collaborative; interdependent. The next breakthrough economy will not be built by isolated brilliance. It will be built by connected capability.

This belief also influences my work with AI ValuePath™, the 10+1 Code™, Devensa, Aigent Connect™, the Visionary Entrepreneur Growth Accelerator©, and Largent Media. These strategic partnerships, although each unique in purpose, share a common objective: helping enterprises transform AI from curiosity into measurable value. Curiosity creates awareness. Execution creates outcomes.

This AI breakthrough belongs to everyone.

Trust: The World's Most Trusted Currency

Recently, I had the privilege of interviewing His Excellency Dr. Mohamed Al Kuwaiti, head of cybersecurity for the UAE government. During our conversation in Dubai, he challenged leaders to think differently about the future.

Think beyond quarterly earnings. Think beyond election cycles. Think beyond 2050. Think beyond 2070. Then he shared a statement that has remained with me ever since: "Trust is the world's most valuable currency."

I agree. In an age of AI, AGI, cybersecurity threats, misinformation, deepfakes, and quantum disruption, trust will become the defining competitive advantage of the 21st century.

Trust is not soft. Trust is infrastructure. Trust is economic capital. Trust is national security. Trust is the bridge between technological possibility and deep and wide economic prosperity. Without trust, adoption slows. Without trust, innovation stalls. Without trust, prosperity becomes fragile. Organizations that build trust will lead from the front.

This AI breakthrough belongs to everyone.

Beyond AI: AGI, Quantum, and Amplified Possibility

Conversations today focus on artificial general intelligence and quantum computing. Predictions vary widely. Notwithstanding, leaders should focus less on speculation and more on preparation.

Imagine accelerating medical discovery. Imagine identifying new treatments in months rather than years. Imagine optimizing energy systems. Imagine reducing supply chain calamities. Imagine unlocking new breakthroughs. Imagine personalizing education at scale. Imagine empowering entrepreneurs with sustainable pathways of EBITDA and unprecedented access to intelligence and strategic insight.

These possibilities represent far more than technological advancement. They represent economic advancement. Human advancement. Societal advancement.

The greatest breakthrough is not artificial intelligence itself. The greatest breakthrough is human potential amplified at scale.

America's Next 250

As we celebrate America's 250th anniversary, we face a defining choice. We can fear disruption. Or we can lead transformation.

We can concentrate on capability among a few. Or we can democratize capability for millions. Not just Silicon Valley. Not just Wall Street. Not just Washington DC. Not just the Fortune 500.

This AI breakthrough belongs to everyone.

The small business owner. The educator. The manufacturer. The farmer. The veteran. The student. The doer.

This AI breakthrough belongs to everyone.

I remain energized and optimistic because I have seen the ingenuity of the American entrepreneur. I have seen innovation emerge from unexpected places. I have observed communities unite around shared purpose. I have witnessed leaders transform challenges into breakthrough economic opportunities.

The next unicorn may emerge from a rural town. A public university. A woman-owned business. A family-owned manufacturer. A tech startup with access to world-class AI capability. The future is bigger than technology. The future is about expanding economic opportunity.

A Call to Build

If AI expands intelligence, we must expand economic opportunity. If AI expands capability, we must expand responsibility. If AI expands productivity, we must expand prosperity and intergenerational wealth for all.

The future will not be built by technology alone. It will be built by people. People with integrity. People with vision. People with courage. People with energy, discipline, and grit. People willing to build what has never been built before and willing to solve what has never been solved. Artificial intelligence may become the most powerful tool humanity has ever created. It is up to us because tools do not shape destiny. People do.

The next breakthrough economy will not be measured solely by productivity gains or technological milestones. It will be measured by our ability to empower stronger institutions, flourishing small businesses, thriving communities, broader economic opportunity, and a more prosperous future for generations we will never meet.

This AI breakthrough belongs to everyone.

The responsibility belongs to all of us. The opportunity is now. America's next 250 years are not waiting to be predicted. They are waiting to be built. With integrity. With trust. With energy. With discipline. With grit. With action. With responsibility. And with an unwavering conviction that the greatest breakthroughs are in front of us. The future is not waiting. And neither should we.

References

1. Anand S. Rao and Gerard Verweij, *Sizing the Prize: What's the Real Value of AI for Your Business and How Can You Capitalise?* (London: PwC, 2017), https://www.pwc.com/gx/en/issues/analytics/assets/pwc-ai-analysis-sizing-the-prize-report.pdf. PwC estimates artificial intelligence could contribute up to $15.7 trillion to the global economy by 2030.
2. Michael Chui et al., *The Economic Potential of Generative AI: The Next Productivity Frontier* (McKinsey & Company, June 14, 2023), https://www.mckinsey.com/capabilities/tech-and-ai/our-insights/the-economic-potential-of-generative-ai-the-next-productivity-frontier. McKinsey estimates generative AI could add between $2.6 trillion and $4.4 trillion annually to the global economy.
3. See OECD, "Real GDP Long-Term Forecast," accessed July 8, 2026, https://www.oecd.org/en/data/indicators/real-gdp-long-term-forecast.html; John Hawksworth, Hannah Audino, and Rob Clarry, *The World in 2050: The Long View: How Will the Global Economic Order Change by 2050?* (London: PwC, February 2017), https://www.pwc.com/gx/en/world-2050/assets/pwc-the-world-in-2050-full-report-feb-2017.pdf; World Bank, *Global Economic Prospects, January 2026*(Washington, DC: World Bank, 2026), https://www.worldbank.org/en/publication/global-economic-prospects. While forecasts vary by methodology and scenario, multiple projections indicate that global GDP could exceed $200 trillion by mid-century.
4. US Small Business Administration, Office of Advocacy, *Frequently Asked Questions About Small Business 2026*(Washington, DC: U.S. Small Business Administration, February 2026), https://advocacy.sba.gov/2026/02/03/frequently-asked-questions-about-small-

business-2026/. There are 36,207,130 small businesses in the United States.

About the Author

Mark L. Madrid is a nationally recognized economic strategist, AI futurist, and founder and CEO of Breakthrough Mavens, LLC, an AI-powered fractional C-Suite boutique firm helping organizations transform vision into measurable results.

As former Associate Administrator of the US Small Business Administration, Mark and his team supported more than four million small businesses through entrepreneurship, capital access, procurement, cybersecurity, and rural economic development initiatives.

A bestselling author and sought-after speaker, Mark authored *ENERGY: 21 Uncomplicated & Easy Tips to Start Your Day Energized by 9 AM* and co-authored *The Leader's Playbook*. He is host of the *Self-Fueling Excellence*™ streaming channel and the *Small Business Is BIG Business with Mark L. Madrid* podcast.

Mark currently serves as entrepreneur-in-residence at NMSDC, senior executive advisor to First National Bank of Bastrop, and visiting scholar at Clinton School of Public Service. An honors graduate of both The University of Texas at Austin and University of Notre Dame, he is an Honorary Colonel in the U.S. Army Reserve and Jefferson Award recipient for his lifelong dedication to public service. He is proud member of the Hero Club and the C-Suite Network.

Website: www.marklmadrid.com

LinkedIn: https://www.linkedin.com/in/markmadrid/

CHAPTER 18

THE AI TRANSFORMATION IS NOT ABOUT AI

By Leonardo Veri
CTO, Tech Advisor, and Founder
São Paulo, Brazil

Art does not reproduce what we see; rather, it makes us see.

—Paul Klee

Every era of transformation gets named after the wrong thing. We called it the Industrial Revolution, but it was not about steam engines. It was about what steam engines made possible: the factory, the city, the supply chain, the modern corporation. We called it the Digital Revolution, but it was not about computers. It was about what computers made possible: instant communication, global markets, the disaggregation of information from physical media.

Now we are calling it the AI Revolution. And we are making the same mistake again. We are naming the transformation after the technology, and in doing so, we are asking the wrong questions, building the wrong strategies, and measuring the wrong outcomes.

The AI transformation is not about AI. It never was. It is about what AI makes possible and what we choose to do with it.

The Adoption Trap

Organizations across every industry are investing billions into artificial intelligence. Boards demand AI strategies. Leaders announce AI initiatives. Entire functions are restructured around automation. And yet, the majority of AI transformation programs fail to produce the business outcomes their sponsors expected.

The technology works. The models are capable. The tools are available and increasingly affordable. What fails is something else entirely.

Most organizations treat AI transformation as a technology adoption program. They procure tools. They run pilots. They hire prompt engineers and data scientists. They build proofs of concept that impress in demonstrations and stall in production. And then they discover what no vendor presentation warned them about: adding intelligence to a confused organization produces intelligent confusion.

You cannot automate a broken process and call it transformation. You cannot deploy AI on top of data you do not trust. You cannot accelerate a strategy that was never clear to begin with. When the foundation is wrong, speed only takes you further in the wrong direction.

The failure is not technological. It is architectural. Organizations are attempting to transform without redesigning. They are installing a new engine into a vehicle built for a different road, then wondering why performance has not improved.

The question is not "How do we adopt AI?" The question is "How do we redesign for leverage?" These are not variations of the same question. They produce entirely different answers.

The Two Operating Systems

In my previous chapter, "Clarity Is the Real AI Opportunity," in *The AI Opportunity* book, I argued that as intelligence becomes abundant, clarity becomes the true source of competitive advantage. AI does not fail at execution. It fails at definition. When problems are vague, AI amplifies vagueness. When strategy is misaligned, AI scales misalignment. The technology is not the bottleneck. Clarity is.

That chapter was about how to think. This one is about how to transform.

Genuine AI transformation requires two complementary operating systems working together. The first is AI First Thinking. This is the cognitive operating system, the discipline of defining problems with precision, mapping organizational reality honestly, designing better decisions, and placing intelligence at the core of how leaders navigate the world. The VeriLabs Compass operationalizes this thinking across five dimensions: True North (Problem Clarity), Reality Mapping (Context and Systems), Decision Architecture, AI at the Core, and Execution and Feedback. The Compass is not a project methodology. It is a leadership philosophy. A way of approaching every challenge with the rigor required to make AI meaningful rather than cosmetic.

The second is AI Native Execution. This is the organizational operating system, the structural redesign that translates clarity into compounding momentum. This is where most strategies break down. Leaders understand the direction but cannot mobilize the system. Plans remain in slide decks. Initiatives remain in committees. The engine never starts. The Flywheel never spins.

The relationship between these two operating systems is the central insight of AI transformation. The Compass gives direction. Without it, organizations move fast in the wrong direction. The Flywheel creates momentum. Without it, organizations never leave the planning stage. Both are required. Neither is sufficient alone.

The VeriLabs Flywheel: An AI Native Growth System

The Flywheel is not a linear methodology. It is a self-reinforcing cycle, an AI Native Growth System that builds momentum through five compounding stages. Like the physical flywheel it draws its metaphor from, it requires significant initial effort. But once spinning, each stage amplifies the next. And unlike a project, it does not end. It becomes the new operating model.

1. Data Foundation: Can We Trust Our Data?

Everything begins here. Before deploying intelligence, an organization must establish a trusted data layer, a single source of truth across CRM, ERP, marketing, finance, operations, and customer interactions. This is not an IT infrastructure project. It is a strategic precondition, the organizational equivalent of paving roads before opening markets.

The defining question of this stage is not "Do we have data?" Almost every organization has vast and growing amounts of data. The question is "Can we trust our data?" Until the answer is yes, every AI initiative is built on a foundation it cannot support. Organizations that skip this stage are not moving faster. They are amplifying their existing errors at machine speed.

2. Revenue Intelligence: Where Is the Money Hiding?

A trusted data layer becomes commercially powerful when it begins to reveal what was previously invisible. Revenue Intelligence transforms raw data into actionable insight: customer behavior patterns, pipeline performance, conversion dynamics, churn signals, and the hidden commercial opportunities that exist in every business but escape notice in the noise of daily operations.

This is the stage that makes AI transformation financially legible. It produces the early wins, concrete, measurable, attributable results that build organizational confidence and justify continued investment in the Flywheel. Revenue intelligence answers the question that senior leadership actually cares about: Where is the value we are currently missing?

3. Process Intelligence: What Should Be Optimized, Automated, or Redesigned?

Once the organization can see its commercial reality clearly, it must map its operational reality with equal honesty. Process Intelligence identifies bottlenecks, inefficiencies, friction points, and high-cost repetitive activities distributed across the operating model.

The critical discipline here is distinguishing among three different responses to what Process Intelligence reveals. Some processes should be optimized, improved within their current logic. Some should be automated, removed from human execution entirely. And some should be redesigned from the ground up, rebuilt around AI-native assumptions rather than the inherited assumptions of the pre-AI era. Most organizations attempt to automate what should have been redesigned. They make their old mistakes faster. Process Intelligence prevents this by clarifying not just what needs to change, but what kind of change each process actually requires.

4. Agentic Operations: What Can Operate Autonomously?

This is the deployment stage. AI agents and intelligent workflows take over routine execution, creating operational leverage and freeing human capacity for the higher-order work that defines organizational advantage: judgment, strategy, creativity, and relationship.

The defining question of this stage is not whether to deploy agents, but where the boundaries of autonomous operation should be set. What should run without human review? What should be human-supervised? What should remain entirely in human hands? These are not technology decisions. They are governance and design decisions, and they require exactly the organizational clarity that the previous stages have built. Organizations that reach this stage having invested in their data foundation, revenue intelligence, and process redesign consistently find that they can delegate far more to autonomous systems than they initially believed.

5. Continuous Growth: How Do We Create a Self-Improving Organization?

The final stage closes the loop. Feedback from agentic operations continuously refines decisions. Better decisions improve processes. Better processes generate higher-quality intelligence. Higher-quality intelligence strengthens the data foundation. Each cycle makes the next cycle more effective.

This is the stage at which the Flywheel becomes self-sustaining, the point at which an organization transitions from running good programs to building a structural compounding mechanism. The result is not temporary efficiency gains. It is durable competitive advantage, the kind that is genuinely difficult to replicate because it is embedded in how the organization thinks, operates, and learns.

The Leverage Transformation

What the Compass and the Flywheel together represent is something I call a Leverage Transformation, a fundamentally different category of change from what most organizations pursue under the label of AI adoption.

Technology adoption is additive. You add tools to existing structures and the structure itself remains the same. Leverage Transformation is multiplicative. You redesign the structure so that every human decision, every business process, every organizational system is amplified by intelligence operating at scale. The former produces efficiency. The latter produces reinvention.

The AI era is already surfacing a profound paradox. Organizations are reaching unprecedented levels of productivity and market valuation. And yet teams are getting smaller. Employment structures are being disrupted. Individuals with AI leverage are outperforming departments. A strategist with the right AI system can analyze markets at a depth that previously required analyst teams. A developer with the right tools can build systems at a speed that previously required engineering departments.

This is not a story about machines replacing humans. It is a story about what happens to the ratio between human effort and human output when AI is involved. That ratio has changed structurally. And the organizations and individuals who understand this are not asking, "How do we survive?" They are asking, "How do we redesign around this new reality?"

As Naval Ravikant has observed, specific knowledge combined with leverage creates wealth that cannot be commoditized.[1] In the AI era, leverage is no longer scarce. What remains scarce and what, therefore, defines the true competitive frontier, is the human capacity to wield it with clarity and judgment.

The capabilities that become most valuable are precisely those that AI cannot replicate: the ability to define the right problem in the right frame; the judgment to make good decisions under uncertainty with incomplete information; the systems thinking to see second- and third-order consequences before they materialize; the creativity to generate genuinely novel ideas rather than sophisticated recombination of the familiar; and the wisdom to know what should remain human. These are not soft skills. They are the highest-leverage human skills. And they compound powerfully when amplified by AI Native systems.

The Question That Remains

At the end of every AI transformation engagement, after the frameworks have been presented, the roadmaps built, the early wins celebrated, there is always a moment of quiet in the room. And in that moment, the most honest leaders ask a version of the same question: "Are we becoming the kind of organization that deserves to win in this era?"

That question is the right one. Not "Have we adopted enough AI?" Not "Are we ahead of our competitors?" But: "Are we redesigning ourselves at the level the moment requires? Are we building people who can think clearly, judge wisely, and orchestrate intelligence? Are we building systems that compound rather than just run?"

The Compass gives direction. The Flywheel creates momentum. But the transformation itself is a choice, a commitment to redesign rather than decorate, to reinvent rather than simply optimize, to become rather than merely adopt.

In a world where intelligence is increasingly abundant and increasingly cheap, the true differentiators are the ones that cannot be automated: clarity of purpose, quality of judgment, and the courage to commit before certainty arrives.

The AI Transformation is not about AI. It is about redesigning individuals, teams, and organizations around leverage. The organizations and individuals who recognize this early, and act on it with both clarity and conviction, will not merely survive the AI era. They will define it.

References

1. Eric Jorgenson, *The Almanack of Naval Ravikant: A Guide to Wealth and Happiness* (Magrathea Publishing, 2020).

About the Author

Leonardo Veri is the founder of VeriLabs, an AI venture studio and strategic consultancy dedicated to helping organizations redesign themselves for the AI era. He works with CEOs, founders, and executive teams to transform AI potential into compounding organizational leverage.

With over two decades of experience across enterprise technology, cloud computing, innovation strategy, and AI transformation, Leonardo has advised organizations ranging from large enterprises to high-growth startups across multiple industries. His work focuses on problem framing, decision architecture, and the design of AI Native operating models.

Through frameworks including the VeriLabs Compass and the VeriLabs Flywheel, he helps leaders move from AI adoption to AI Transformation, from thinking AI First to executing AI Native.

Leonardo is also a keynote speaker, global mentor, and host of the VeriTalks podcast.

Email: leonardo.veri@gmail.com

Website: www.leoveri.com

LinkedIn: https://www.linkedin.com/in/leonardoveri/

CHAPTER 19

HOW AI WILL MAKE US MORE HUMAN

By Nicolas Mendiharat
Serial Founder and CEO
New York, New York

It is not that we have a short time to live,
but that we waste a lot of it.

—Seneca

Every civilization is shaped by a few inventions that are not merely useful, but irreversible. When being awed or afraid of AI, we often forget history's perspective:

The wheel changed trade, war, agriculture, and distance itself. Agriculture created settlement, surplus, cities, hierarchy, taxation, and even writing. The printing press reorganized knowledge and authority. Electricity stretched the productive day and powered modern urban life. Computers created an entire digital layer on top of reality.

Artificial intelligence belongs in this family. But it is different from the great technological leaps before it in at least two decisive ways: speed and freedom. The speed is different because AI develops at computer speed, not human speed. The printing press spread across Europe over decades. Electricity took generations to build into streets, homes, factories, and machines. Even the internet unfolded in waves. AI is moving faster because the tool is helping build the next version of the tool.

The freedom is different because, for the first time, the central promise of technology may not be that humans can work harder, longer, or at greater scale. The promise may be that humans can stop doing much of the work that keeps them trapped in the rat race. Previous technologies often increased output but did not necessarily return time. The smartphone made everyone permanently reachable. The computer made office work more productive, but also created more spreadsheets, more emails, more dashboards, and more expectations.

AI could become another version of that trap of course, and that's part of the risks. But it also opens a more radical possibility: that intelligence at scale can give us back the hours, attention, and energy that make us human in the first place.

To understand why AI can make us more human, we first need to understand why it will not replace us.

Why AI Will Not Replace Humans
The Category Error

The most common fear about AI is that it will replace human beings. It will certainly replace tasks, roles, and business models. Anyone who denies this is not paying attention. But replacing work is not the same as replacing humans.

The mistake comes from confusing intelligence with humanity. AI can already write, code, translate, create images, analyze contracts, structure data, and answer questions at a level that would have seemed impossible a few years ago. But human beings are not question-answering machines. We are embodied, emotional, social, contradictory, conscious, hormonal, mortal creatures moving through a physical world.

David Hume and the empiricist tradition have never been more on point in that regard. They remind us that knowledge begins in experience. We do not encounter the world as clean abstractions. We feel heat on skin, smell rain on pavement, hear the tone in a friend's voice, and taste a wine differently depending on the room, the memory, and the person across the table. A machine can process sensory data, but it does not live inside the senses.

Then come emotions. Love is not a prompt. Desire is not a data structure. Grief is not a statistical distribution. We do not merely think; we are moved.

Consciousness is even more difficult. Philosophers have debated it for centuries, neuroscientists still struggle to explain it, and technologists are often too quick to reduce it to computation. A system may simulate a conversation about consciousness. That does not mean it has a first-person experience of being alive.

There are also simple human pleasures that reveal the gap. AI can analyze a tennis swing, optimize a runner's training plan, or recommend a mountain trail. But it cannot enjoy the exhaustion after a 10K run, the quiet satisfaction of physical progress, or the irrational joy of being outside on a perfect day. It cannot play sports for pleasure because pleasure is not an output. It is a lived state.

Humor may be one of the clearest examples. AI can generate jokes, and sometimes they are technically amusing. But stand-up comedy is not only joke construction. It is timing, risk, room energy, body language, and the strange human ability to turn pain into laughter. A machine may imitate the structure. It does not feel the danger of the silence before the laugh.

Finally, there is free will. Some people believe free will is an illusion, that everything is determined by biology, environment, God, fate, or physics. If so, the debate about AI replacing human agency becomes less interesting, because human agency was already an illusion. But if free will exists, even imperfectly, then it is central to what makes us human. We choose what to value, whom to love, what to pursue, and what kind of life to build.

The Human Pushback

Recent history has already corrected some of our technological fantasies. The metaverse was supposed to become the next great environment for human life. Meta invested tens of billions of dollars through Reality Labs, yet the broad social migration into virtual reality did not happen as expected.[1] Most people do not want to spend their lives as avatars in headsets. They want dinner, sunlight, eye contact, travel, sport, flirtation, serendipity, and physical presence.

The pandemic produced a similar misconception. Many predicted that conferences, fairs, concerts, and large gatherings would never fully return. Zoom was cheaper and more practical. But "well enough" is not the same as human. By 2025, live entertainment had reached record scale, with Live Nation reporting 159 million fans attending its events worldwide.[2]

I see the same instinct in smaller social signals. In meetings and restaurant dinners, people seem to be putting their phones away more than they did a few years ago. Not always, of course, and not everywhere. But the gesture matters. It is as if intelligent humans are starting to build their own safeguards before any regulator or platform does it for them.

At the TED main conference in April 2026 in Vancouver, I watched a poet begin her talk by saying that her talk was not about AI, and the room gave her a standing ovation before she had really begun. Around the same period, I heard AI experts being booed at a technology conference while praising AI too uncritically.

These moments are anecdotal, but they are not trivial. They suggest that the most sophisticated audiences are not rejecting technology; they are defending the human space around it.

New York real estate tells a related story. After 2020, many announced the death of cities, offices, density, and urban life. But human beings still want proximity, culture, romance, restaurants, randomness, and the electricity of other people. And now relative real estate prices in New York are as high as ever.

Even at the level of computation, the human brain remains extraordinary. It operates on roughly the energy of a small light bulb while coordinating perception, movement, memory, emotion, language, and social intelligence. The often-cited figure is around twenty watts for the brain.[3] Computers will advance, but replicating the whole human package is not simply a matter of more chips.

AI will replace many things humans do. It will not replace being human.

Why AI Will Make Us More Human
The Return of Time

The optimistic case for AI is not that it will make us more productive. Productivity is too small a word. The deeper possibility is that AI will help us reclaim time.

Time is the hidden currency of every technological revolution. We say we want money, status, efficiency, growth, scale, and convenience. But underneath all of it, we want time: time to think, move, love, learn, create, travel, sleep, heal, and enjoy the short miracle of being alive. Seneca understood this two thousand years ago. We are not poor in time; we are careless with it.

The modern economy has trained us to convert every efficiency gain into more work. If email is faster than letters, we send more emails. If smartphones allow work anywhere, work follows us everywhere. The tragedy of digital productivity is that it often accelerates the machine without liberating the person inside it. AI gives us the chance to break that pattern, but only if we choose to use it that way.

The French thirty-five-hour workweek is a useful, if controversial, example. It was never a simple miracle policy, but it showed that reducing time at work does not automatically mean reducing output. It can force better organization and a sharper distinction between hours spent and value created.[4] The same intuition made Tim Ferriss's *The 4-Hour Workweek* a global bestseller: Eliminate low-value tasks, automate what can be automated, delegate what should be delegated, and design life around freedom rather than busyness.

AI makes that philosophy available at a completely different scale. It can summarize research, draft documents, build models, prepare presentations, compare contracts, translate conversations, organize travel, coach learning, and automate administrative work. The goal is not to become lazy. The goal is to stop confusing friction with virtue.

This is not theoretical for me anymore. Over the last months, I started setting up Perplexity Computer and Gemini NotebookLM for complex and automated tasks, and of course Claude does miracles for running our financial models at our firm. Better tools will exist by the time this book is printed, which is precisely the point. In March, I finished in a couple of hours what I had expected would take me two days. That afternoon, instead of staying in front of a screen to prove my seriousness (not that I think I need that anymore, but for the purpose of this example), I went skiing.

That day captured the whole argument of this chapter. The ideas and structure here are mine. AI did not replace the thought, the conviction, or the lived experience behind them. But it helped me organize the material better, faster, and with less friction. The human part was the point of view and the judgment. The machine helped clear the road.

The End of Screen Captivity

The most important example may be screen time.

Since the 1950s, many consumer technology jumps have pulled human attention toward screens. Television brought the screen into the living room. Personal computers brought it to the desk. Smartphones put it in the pocket. Social media turned the screen into a casino of attention. Notifications turned every moment into an interruption.

The result is that technology has swallowed an enormous share of waking life. We work, relax, date, shop, learn, navigate, and often fall asleep through screens. Studies associate heavy screen time with poorer sleep and mental-health outcomes, especially among younger populations.[5] Screens are useful. But a life mediated primarily through glass is smaller than one lived through the body, the senses, and other people.

AI can reverse this trend through three cumulative effects. First, computers can execute tasks much faster. For decades, we adapted ourselves to computer workflows: menus, files, tabs, formulas, dashboards, and endless interface management. AI collapses much of that distance. We can tell the machine the outcome we want, review the result, refine the judgment, and move on. The human role shifts from operator to director.

Second, voice interaction with large language models is becoming precise enough to move computation away from the screen. When we can speak naturally to machines, the physical posture of digital life changes. We can walk, cook, drive, exercise, or think out loud while interacting with intelligence. As AI glasses and other contextual devices become mainstream, the interface can become less like a portal we disappear into and more like an ambient assistant that accompanies the real world. The important direction is clear: less staring, more living. Alexa was the entertainment command version of this; your smartphone becomes your voice-does-everything command.

Third, AI will make humans better learners. Education has always been limited by standardization. A teacher with thirty students cannot customize every explanation in real time. A textbook cannot know what you misunderstood. AI can become a tutor, translator, coach, debate partner, editor, simulator, and memory system. It can explain quantum physics through skiing, negotiation through dating, or tax law through real estate.

This matters because learning is one of the most human activities per se. We become more ourselves by learning: a language, a sport, an instrument, a craft, a philosophy, a market, a city, another person. AI can compress the boring parts of learning and expand the meaningful parts.

If we use it well, AI will not make us passive. It will make us more capable. It can reduce the time spent formatting, searching, and coordinating so we can spend more time thinking, understanding, moving, resting, and being with people. This is the real promise: not artificial intelligence instead of human life, but artificial intelligence in service of human life.

The Wrong Companion

There are limits, of course. I am not interested here in the most dystopian scenarios: machines taking over the world, governments using AI to control citizens, corporations predicting every desire. Those risks matter, but they are not the point of this chapter.

The more immediate question is how AI will make less human those who let AI control them. My biggest concern here is mental health.

This is not new. Humans have always surrendered agency to systems that promise certainty, pleasure, belonging, distraction, or power. Religions, for instance, can elevate the soul, but they can also suppress independent thought. Propaganda can organize nations around lies. Television can pacify attention. Social media can engineer comparison, outrage, and addiction.

AI will make those "machines" more intelligent. If someone decides, consciously or unconsciously, to spend eight hours a day in algorithmic entertainment, AI will make that entertainment more personalized and more difficult to leave. If someone wants to avoid thinking, AI will provide ready-made opinions. If someone wants confirmation, AI will produce it. If someone wants distraction, AI will never run out.

This is where AI companions worry me. An animal companion taps into genuine human empathy. A dog or a cat is not a substitute for a person, but the bond is embodied and grounded in care.

I understand that talking to a bot may help some people, especially in moments of loneliness. But I have not yet met someone who seems happier because they replaced human connection with a chatbot. The risk is not that the bot is unintelligent. The risk is that it becomes intelligent enough to make isolation feel acceptable.

This brings us back to free will. AI will not make people more human if they have already decided to allocate very little time to human things. It will not make a passive person active by magic. It will not replace discipline, education, taste, judgment, or courage.

In this sense, AI is less like a destiny than a mirror. It magnifies the user's intention. If your intention is freedom, AI can help you reclaim time. If your intention is avoidance, AI can help you avoid life more efficiently. If your intention is creation, it can multiply your output. If your intention is consumption, it can deepen the feed. The question is not only what AI will do to us. The question is what we will ask AI to do for us.

Conclusion: The Choice Before the Task

The more you want to reclaim your time to do human things, the more you should understand AI and take advantage of it. Ignoring it will not protect your humanity. Fear will not protect it either. The only protection is mastery.

Mastery begins with discomfort. It takes effort to learn a new tool, change habits, question old workflows, and admit that many tasks we once considered important were merely familiar. But this is exactly what philosophy, and Buddhism as a spiritual example, have always taught in different languages: Do not run from discomfort; Observe it, enter it, understand it, and transform it into freedom.

The way I use AI today has already forced me to unlearn. Before I start a task, I try not to lean immediately on experience, instinct, or the old way of doing it. I first ask a more useful question: Can AI save me time here, and can I prompt it through voice or text before I open another screen, another tab, another spreadsheet? That small pause turns AI from a toy or threat into leverage.

Epictetus wrote that some things are up to us and some are not. AI's existence is no longer up to us. Its development will continue. What remains up to us is how we relate to it. Do we become servants of the machine, or do we make the machine serve the parts of life that machines cannot live?

AI will not make us more human automatically. But it can make us more human if we use it to recover what industrial and digital life have taken from us: time, attention, presence, movement, learning, conversation, and choice.

The future will not belong to people who reject AI. It will not belong to people who worship it either. It will belong to people who understand it well enough to put it in its place. And its place is not above us. Its place is behind us, beside us, and beneath the human life we still have to choose.

References

1. Felix Richter, "Meta's Reality Labs Lose $19B as VR Ambitions Falter," *Statista*, January 29, 2026, https://www.statista.com/chart/29236/operating-loss-of-metas-reality-labs-division/.
2. Live Nation Entertainment, "Live Nation Entertainment Full Year and Fourth Quarter 2025 Results," press release, February 19, 2026, https://newsroom.livenation.com/news/live-nation-entertainment-full-year-and-fourth-quarter-2025-results/.
3. William B. Levy and Victoria G. Calvert, "Communication Consumes 35 Times More Energy than Computation in the Human Cortex, but Both Costs Are Needed to Predict Synapse Number," *Proceedings of the National Academy of*

Sciences 118, no. 18 (May 4, 2021): e2008173118, https://doi.org/10.1073/pnas.2008173118.

4. Gunnar Trumbull, "France's 35 Hour Work Week: Flexibility Through Regulation," *Brookings Institution*, January 1, 2001, https://www.brookings.edu/articles/frances-35-hour-work-week-flexibility-through-regulation/.
5. Benjamin Zablotsky et al., "Associations Between Screen Time Use and Health Outcomes Among US Teenagers," *Preventing Chronic Disease* 22 (July 10, 2025): 240537, https://doi.org/10.5888/pcd22.240537.

About the Author

Nicolas Mendiharat is a serial entrepreneur who has founded and led seven consumer tech startups built around his lifestyle and passions. He is now channeling that experience into a new private equity firm, Permanium Capital. He began with a ski tour operating company for students at 22, selling it five years later to leading French travel group Nouvelles Frontières. He then developed Busresa, a profitable bus transportation software from day one, which grew into its own startup and was sold in 2004.

In 1999, Nicolas launched skihorizon.com, which became a leading ski vacation booking site with 300,000 customers and nearly $100 million in annual sales. He later founded Palate Club, an AI-driven wine taste-matching platform, and co-developed a $25 million four-star residence in Les 3 Vallées. He also co-founded Winechain, a blockchain-based fine wine marketplace, before creating Permanium Capital, a hybrid private equity platform focused on ultra-luxury services, real estate, and hospitality.

Email: nmendiharat@gmail.com

Website: www.permanium.com

LinkedIn: https://www.linkedin.com/in/mendiharat/

CHAPTER 20

ARCHITECTING INTELLIGENCE: SCALING AGENTIC AI AND AIOPS FOR ENTERPRISE IMPACT

By Yashaswini Nalla
AI Technology Leader
Collierville, Tennessee

Imagine you are in the meeting. A senior leader or more likely, a face in a grid of video tiles says, "We need to scale AI this year."

The architect unmutes, "We don't have guardrails yet."

The operations lead follows, "Who gets paged if the system makes the wrong call?"

Silence.

Not the silence of confusion, the silence of people who understand technology but have not resolved what comes after deployment. Nobody in that room doubts the model anymore that debate is over. What is still unresolved is far more uncomfortable: Who owns decisions once systems start making them? This is what AI transformation looks like inside organizations today. Not excitement or fear but tension. Forward momentum constrained by unresolved ownership not because the technology is not ready, but because autonomy arrived faster than a shared understanding of how to govern it. That silence is not isolated. It is playing out across enterprises on a scale.

McKinsey's 2024 State of AI report found that 72% of enterprises have AI systems in production but only 9% describe their AI governance as mature Boston Consulting Group reported that 74% of companies have yet to see tangible value from their AI initiatives, despite years of experimentation. The gap is not in the models. It is in the organizational architecture surrounding them.

Enterprise technology leaders are rewarded for how quickly AI is deployed, not how responsibly it is governed. Shipping AI features move quarterly metrics; building governance infrastructure does not. At the same time, workforce AI adoption jumped up in the last couple of years, a massive behavioral shift. Employees have embraced AI as a core part of how they work, often without formal approval or policy frameworks in place. This gap between usage and governance is where most enterprise risk now lives. Most organizations using generative AI have already experienced issues like hallucinated outputs, security incidents, privacy exposure, and intellectual property leakage. These are no longer theoretical concerns; they are operational realities.

What that meeting is really exposing is not a technology problem. It is an architectural and organizational reckoning, one most enterprises are only half-prepared for. The real differentiator is not which AI platform a company selects. It is whether the enterprise builds trust, governance, and shared ownership into its AI systems from day one.

That is what this chapter is about. Because the enterprises that lead this decade will not be the ones who deploy AI fastest. They will be the ones who build it right.

The Architectural Shift to Agentic Systems

For years, enterprise technology did exactly what it was told: Follow this rule, execute this workflow, process this transaction, and when something broke, someone knew where to look. The failure was predictable; that predictability was the whole point.

Agentic AI changes that deal entirely.

The shift did not arrive with a big announcement. It crept in through productivity tools, AI assistants, POCs, and pilot programs that quietly expanded in scope. What is happening inside organizations now is not an upgrade to automation, it is a different kind of system altogether. Automated systems follow scripts, but agents pursue goals. They decide what action to take, adapt when circumstances change, and influence outcomes across workflows no one explicitly mapped for them.

Here is a straightforward way to see the difference. A traditional bot in a procurement workflow checks whether an invoice matches a purchase order. It applies to a rule. An agent does something else entirely. It checks patterns, flags anomalies, cross-references vendor history, starts a verification step, and escalates without being asked. Same domain. Completely different relationship between the system and the decision.

Now pause and sit with an uncomfortable question: If that agent makes a bad call today? If it approves the wrong payment? If the agent triggers the wrong action? If it makes a decision that costs the business, do you know who is responsible? Can you explain why it decided what it did? Most organizations cannot, and that is not a technology gap. It is a design gap.

The problem is not that enterprises moved too fast into agentic AI. Many organizations moved with real ambition and tangible results. The problem is that they brought an old mental model with them. They governed agents the same way they governed software, with the assumption that failures are traceable, scope is fixed, and accountability sits in the code, but agents do not fail like software.

They fail, like people, through imperfect judgment and organizations are not yet designed to manage that.

The architectural mindset this demands is straightforward to say, even if it is hard to stop designing AI systems around data flows and start designing them around accountability flows. Who authorized this action? Under what conditions? What happens when it is wrong? These are not technical questions. They are governance questions and they need answers before deployment, not after an incident forces the issue.

The organizations pulling ahead are not always the ones with the most sophisticated models. They are the ones who asked the accountability questions first. They drew clear lines around what agents can decide independently, what needs a human in the loop, and what should never be delegated at all. They built those boundaries into the system from day one, not as a compliance checkbox, but as the foundation the entire system stands on. That is the architectural shift.

Designing Autonomy That Holds at Scale

Most organizations treat autonomy like a light switch: on or off, allowed or restricted. That binary thinking is precisely where scaling falls apart. Autonomy is not a feature you toggle. It is a spectrum you design. The enterprises getting this right are doing three things differently:

- Designing graduated autonomy instead of full delegation
- Setting clear trust boundaries instead of vague policies
- Building defined escalation paths instead of reacting in panic when something goes wrong.

Autonomy Has to Be Earned

Think about how a new employee earns trust inside an organization. They do not walk in on day one with authority to approve contracts or reallocate budgets. They start narrow, prove judgment, and expand their remit as trust is established. AI agents need the same logic applied to them. A practical design runs four levels:

1. Agent simply informing a human
2. Agent assisting and staging actions for review
3. Agent acting within guardrails with exception escalation
4. Agent acting autonomously with periodic audit

The mistake most organizations make is jumping straight to level four in the name of efficiency, before building the oversight infrastructure that makes level four safe. Today, very few enterprises say they are comfortable with fully autonomous AI agents. That is not timidity; it is a rational response to systems that have not yet earned broader trust.

What a Trust Boundary Actually Looks Like

A trust boundary is not a policy document. It is a set of conditions encoded in the system that define exactly where autonomous action stops and human judgment begins.

In financial services, a leading institution built an AI agent to handle fraud triage. The agent could autonomously flag and hold transactions below a defined threshold. Above that threshold, it staged the action and needed analyst approval. When a false positive occurred, the analyst could see exactly what the agent saw, why it flagged the case, and what it would have done. That transparency was not accidental. It was designed in from day one. That is what a trust boundary looks like, not "the AI handles fraud," but "the AI handles fraud under these specific conditions, and here is exactly what happens outside them."

In healthcare, the same logic applies differently. AI systems that analyze diagnostic images and flag potential abnormalities are now common. But leading deployments have one thing in common: The agent does not deliver a diagnosis. It delivers a recommendation with a confidence level, and a clinician confirms before anything reaches the patient. The agent earns more latitude over time as its record builds. That is graduated autonomy working as designed.

Resistance Is Rational and Architecture Resolves It

What most AI transformation programs get wrong is that they treat employee resistance as a culture problem to manage: run training, communicate the vision, and show the ROI. When people cannot explain why a system made a decision, they do not trust it. When they cannot predict how it will behave, they work around it. When they cannot intervene confidently, they disengage. These are sensible responses to systems not designed with the people who use them in mind. The implication for design is direct: Explainability is not a nice-to-have feature for compliance teams. It is the primary mechanism through which frontline employees build the confidence to use and rely on the system. An agent that shows its reasoning, even in plain language, is not just more auditable. It is more adaptable. Trust is not something you request from your users. It is something your architecture earns, interaction by interaction, over time.

The Escalation Path Is Not a Fallback, It Is the Design

One of the most underbuilt parts of any agentic system is the escalation path. Most teams treat it as a fallback, but the best teams treat it as a core design element from day one. A well-designed escalation path answers three questions: What triggers it? Who receives it? And does that person have enough context to act?

That last question is where most implementations fail. The agent escalates, a notification arrives, and the human opens it to find a status flag with no context, no reasoning, no options, and no clarity on what the agent was trying to do. That is not a human in the loop. That is a human receiving a handoff they cannot act on. The organizations getting this right give humans the full picture when escalation happens. That turns escalation from a disruption into a conversation, and that conversation is precisely what responsible autonomy looks like at scale.

How Human Roles Evolve and Why You Must Design for It

There is a quiet tension building inside most organizations right now. As AI systems take on execution, humans are left with something unexpected: not relief, but ambiguity, as in: *"If the system manages the analysis, the triage, the first draft, the classification, where do I fit?"* This ambiguity is why many teams cling to manual control. Not out of fear of AI. Out of fear of being removed from decision-making entirely. That distinction matters enormously for how leaders respond to it.

Organizations that navigate this well do something intentional: They redesign human roles before deployment, not after resistance surfaces. They stop describing what AI will do and start defining what humans will own. The roles that emerge are not lesser versions of what existed before. They are more consequential, judgment ownership, supervision, policy definition, and ethical arbitration but these roles do not appear organically. They must be architected, just like the systems themselves.

What This Looks Like When You Get It Wrong

Here is a situation many teams have lived through, including teams I have worked alongside. A model clears every internal bar: strong metrics, sound business case, ready for pilot. Then something unexpected happens. The team whose work most directly overlaps with the model goes quiet. Deadlines slip. Cooperation stalls. No one flags a technical problem, because the blocker is not technical.

This is exactly what happened in one AI deployment. A model designed to accelerate a core analytical workflow was ready for production pilot. The team that had been doing that work manually and doing it well became an unexpected obstacle. Not because the model was wrong, but because nobody had clearly told them what their role would be after it launched.

They assumed replacement. The reality was the opposite. The model required human evaluators to review outputs, catch hallucinations, validate edge cases, and make judgment calls the model could not. The human role was not disappearing. It was shifting from doing the analysis to owning the quality of it. But that distinction was never communicated proactively. It surfaced only after resistance forced a pause. The team then spent weeks on education and reassurance, conversations that should have happened before the pilot ever stalled. The delay was not a technology problem. It was a design omission.

Design the Role Before You Deploy the System

The lesson is straightforward but consistently skipped: If you want humans to step confidently into judgment and oversight roles, you must define those roles with clarity, with training, and with enough lead time before the system goes live. AI does not remove humans from the loop. It asks them to step into more meaningful loops. Supervision over outputs that carry real consequence. Policy ownership over what the system is allowed to do. Escalation authority when the agent reaches the boundaries of its autonomy.

Organizations that design this transition deliberately through role definitions, training programs, and honest conversations about what changes and what does not move faster and meet far less friction. Those that assume the transition will happen naturally pay for it later, one stalled pilot at a time.

Avoiding Predictable Scaling Failure Modes

Scaling AI does not fail dramatically. It fails quietly through accumulation. One agent becomes ten (one for logs, one for metrics, one for incidents, and one for remediation). Ten becomes fifty. Each was approved for good reasons. Each solved a real problem. But nobody maps how they interact, what they share, or who owns them when something goes wrong.

This is agent sprawl and it is one of the most common ways enterprise AI programs lose control. Not through a single catastrophic failure, but through a growing system that no one fully understands. Decision conflicts appear. Ownership becomes unclear. Accountability diffuses.

Tool fragmentation follows the same pattern. Agents are granted access to APIs and systems opportunistically, one integration at a time. Over months, no single team keeps a complete picture of what each agent can access, what permissions it holds, or what actions it can trigger. Security gaps don't appear suddenly. They accumulate, quietly, until they surface as incidents.

Then comes unmanaged autonomy drift. Agents that began with narrow, well-defined scope gradually absorb more through expanded prompts, new use cases, or simply because teams begin to trust them and ask for more. Nobody explicitly expanded the mandate. It expanded incrementally and unnoticed. The common thread across all three failure modes is the same: the absence of governance infrastructure. Not absence of intent, most teams are aware of the risks, but absence of the structures that make governance operational.

That means:

- Agent registries that track each deployed agent: its scope, its tool access, and its accountable owner
- Change control for how agents evolve over time
- Audit trails that humans can interpret, not just store

The fix is not complexity. It is discipline applied early. The organizations that avoid these failure modes are not more cautious, they are more deliberate. They treat their AI agent ecosystem the way mature engineering organizations treat production infrastructure: with ownership, observability, and accountability built in from the start.

Closing: Build It Right

Go back to that meeting room. The silence that followed three straightforward questions has not changed. What has changed is whether your organization now has answers to them. That silence is not a lack of understanding. It is a gap between deployment speed and organizational readiness and it exists inside nearly every enterprise navigating this shift today.

The organizations that lead this decade will not be defined by how fast they deployed. They will be defined by the decisions they made before deployment about autonomy, accountability, and who owns what when something goes wrong.

- They treated autonomy as something earned, not assumed.
- They designed trust boundaries before incidents forced the issue.
- They redefined human roles before resistance slowed them down.
- They governed their agent ecosystem before sprawl made it unmanageable.

None of this requires slowing down. It requires building differently and building it right. These are not implementation details to revisit after something breaks. They are the foundation that determines whether AI becomes a compounding advantage or a compounding liability.

At this point, most readers will recognize their own organization somewhere in this journey. That recognition matters because it defines what comes next. What does thriving actually look like?

For individuals, the shift is not subtle:

- From execution to judgment
- From consuming outputs to understanding reasoning
- From using tools to supervising systems

The advantage will not belong to those who use AI the most. It will belong to those who understand how it thinks and where it fails.

For organizations, the shift is even more fundamental. AI is not a collection of tools. It is a compounding capability. Every decision you make about autonomy, accountability, trust, and human alignment either accelerates that compounding or constrains it. Every boundary you define, every role you redesign, every escalation path you build with care stacks and shapes the system that everything else depends on.

AI transformation is not a technology upgrade. It never was. It is an architectural and organizational shift and the enterprises that recognize that early will not just adopt AI successfully. They will operate differently because of it.

And the silence in that meeting room? It is not the end of the conversation. It is the moment when the right questions finally start getting asked and where the organizations that build it right begin to pull ahead.

About the Author

Yashaswini Nalla is an AI enterprise architect focused on designing intelligent, scalable systems that drive operational excellence. She holds a Master's degree in Computer and Information Systems from Kent State University, where she also served as a teaching assistant, an experience that shaped her ability to simplify complex systems. Over the years, Yashaswini has built a reputation for transforming systems in measurable ways: reducing manual effort, improving reliability, and advancing how platforms operate at scale. In recent years, she has turned that same lens toward artificial intelligence working at the intersection of operations and architecture. She has led AI education initiatives, pursued focused learning across machine learning and generative AI, and applied these capabilities in real enterprise environments. It is these hands-on experiences including friction, resistance, and breakthrough moments that inform her perspective. Yashaswini writes not as a theorist, but as a practitioner who has sat in meetings, navigated the organizational tensions, and built the systems this chapter describes.

Email: ynalla.tech@gmail.com

LinkedIn: www.linkedin.com/in/yashaswininalla

CHAPTER 21

AI-DRIVEN INFLUENCE: SCALING TRUST, CONTENT, AND COMMUNITY IN THE DIGITAL AGE

By Jeanette Okwu
AI Evangelist in the Creator Economy
Berlin, Germany

Same as it ever was.

—David Byrne

Before we get into artificial intelligence, let's get clear on the term "influencer marketing." This is when brands work with people who have built their own audiences. Sometimes those audiences are huge, but more often they're small, focused, and loyal. Think of a skincare brand teaming up with a dermatologist on Instagram, or a bank working with a finance creator on TikTok. Brands reach people they couldn't reach on their own. Creators get paid to do what they're already doing. Audiences get advice from someone they trust.

People buy from people, not brands. This is even truer in today's kinetic, ad-saturated reality than ever.

For a long time, most marketing teams marginalized this work. Nobody quite knew how to categorize it. Was it public relations because it involved addressing prospects? Was it media buying, because money was changing hands for distribution? Was it brand marketing, because it was meant to shape how people felt about a product? The answer, depending on the company and the year, was usually all three and none of them. Influencer marketing is a young discipline. Like any young discipline, it spent most of its early life shifting between departments and reporting lines. But the channel grew faster than any team could professionalize it.

Now, creators take up almost a quarter of social media budgets, more than any other line item. The creator economy is on track to hit $200 billion by 2026. Most marketers plan to spend even more on creators next year, as TV and print continue to lose ground.

It's worth stopping to ask why this shift is happening. People have spent twenty years learning to tune out ads. Banner ads, pre-roll videos, polished TV spots just don't work like they used to.

At the same time, platforms that used to be brand territory are now run by regular people making content for each other. Half of all media time goes to content made by individuals, not brands. Seventy percent of Instagram engagement comes from accounts people don't even follow. Trust didn't disappear. It just shifted from brands to people.

But growing budgets conceal some discomfort beneath the surface. Marketing leaders say they cannot accurately measure the return on their creator content production spending. They believe the channel works, but intuition does not satisfy a CFO. This is where credible measurement must step up. AI-powered attribution models can now connect creator content directly to conversions and revenue, even across multiple platforms and touchpoints. Real-time sentiment analysis, audience engagement scoring, and advanced tracking pixels give a far more precise view of what is actually moving the business. Tools like AI-based influencer analytics dashboards or cross-platform

ROI trackers can help teams build a measurement pipeline that their finance department will recognize.

Content Is Cheap Now. Trust Still Isn't

The debate has devolved into one question: Will AI replace creators? But asking this keeps everyone looking in the wrong direction.

Yes, machines can churn out endless content for almost nothing. However, audiences don't respond to content alone. They respond to the person behind the message: the dermatologist whose solid advice has guided them for years on Instagram. The developer on YouTube whose tutorials saved them an afternoon. The finance creator on X who answered their question in the comments at 11 p.m.

AI cannot manufacture that kind of relationship. What it might do is flood the feed with more of everything else, and the more synthetic the feed becomes, the more valuable a trusted human voice gets.

What AI can do, and this is where its dual impact becomes clear, is split into two roles, depending on the user. For creators, AI acts as a productivity booster. It develops video scripts, generates voiceovers, and can turn a single idea into versions for multiple platforms. This enables creators to scale content production efficiently, freeing up more time to build audience trust. For marketing teams, AI acts as an operational engine. It analyzes data at scale to identify which creators fit a brand, supports briefing and reporting, and conducts real-time sentiment analysis of audience reactions. Each group benefits uniquely: Creators unlock creative potential and save time, while marketers gain tools to strategically manage and measure the channel.

For them, AI takes on the operational weight that has kept creator marketing buried in spreadsheets for a decade. It analyzes data at scale to identify which creators fit a brand. It conducts real-time sentiment analysis of audience reactions. The two uses look similar from the outside but solve different problems. One frees the creator to make better work. The other lets the brand finally manage that work as a real channel.

When both sides of that picture work together, something larger begins to unfold. Influencer marketing stops being a series of one-off campaigns held together by goodwill. It starts looking like a business function you can plan, measure, and rely on. It becomes architectural, not improvisational. That shift is the real story of what AI is doing to this industry.

Part One: Fix the Plumbing

For most of the last decade, creator marketing has run on a strange mix of guesswork and grunt work. Briefs get sent, rewritten, and sent again. Content lands across dozens of platforms with no consistent way to know what drove what. Reports get cobbled together by hand the night before the client calls. None of this is anyone's fault. The tools to do better simply did not exist at the scale the work demanded.

AI changes the math for every step that used to slow things down. Discovery, briefing, production, and reporting stop being human bottlenecks. The piece worth dwelling on is measurement. Measurement has been the industry's Achilles' heel for a decade. Influencer marketing has long been the line item nobody could defend. It is present in every budget and haunts every quarterly review. AI now enables the connection of creator content to commercial outcomes. It does this across platforms and audiences in a way that stands up to scrutiny.

The Operational Drag Is What Gets Replaced, Not the Creator

None of this replaces the creator. The judgment, the taste, the relationship with the audience, the cultural fluency that tells a person which joke to make and which to skip, none of that comes out of a model. What gets replaced is the operational drag. These are the hours spent copying numbers from one platform to another. These are the late nights chasing missing assets. The follow-up emails about contract clauses.

AI does the best work nobody wanted to be doing in the first place. When that work disappears, the people involved can spend their time on the parts of the job that actually move the business: strategy, relationships, creative judgment. The is the human layer that will make all the difference and is hard to replace.

When the Plumbing Works, the Function Becomes Real

Here is what architectural thinking looks like in practice. When discovery, briefing, production, and measurement all run through systems that talk to each other, creator marketing stops being a collection of campaigns. Now it can be a function planned against business objectives, just as paid media is. It can be budgeted with confidence. It can be reported in a language a CFO understands. They don't care about reach or engagement. All they care about is whether a campaign contributed to the bottom line. In their book, this means conversions and revenue. Teams that get this running stop defending their existence every quarter. They start being asked how to scale it.

To start this transition, a few initial steps make a big difference. First, audit your current creator marketing processes to identify manual bottlenecks and areas lacking integration. Then, choose a pilot program to test AI-powered tools that connect discovery, content management, and measurement. Involve cross-functional partners early so everyone understands what operational change will look like. Set baseline metrics at the start, and track improvements as the system takes hold. Small wins build momentum and help make the case for broader adoption.

Part Two: Thought Leadership Is Now a Performance Channel

The second thing AI changes is what counts as authority. As the feed fills up with chatbot-generated content, expertise tied to a real person with a real track record becomes scarce. A polished AI essay carries no weight. A working practitioner with a community, a point of view, and valuable insights outweighs it by a landslide. So brands should think about who they put visibly on stage.

Traditionally, thought leadership is stuffed in the brand-awareness corner of the comms budget. A founder gives a keynote. A CMO writes an opinion piece. A product expert sits on a panel. It looks impressive, gets logged as PR, and nobody really expects it to drive revenue.

The same forces that make creator marketing more valuable are also turning expert content into a real performance channel. When audiences trust people more than brands, the most valuable thing a brand can do is associate itself with the right people in the right conversations.

Generative Search Changes the Discoverability Game

Generative search makes this more urgent than most have absorbed. A growing share of the population now starts their information search with an AI assistant rather than a traditional search engine. AI-generated answers already appear above the regular results on search engines. These answers are not built from paid search slots or sponsored posts. They are stitched together from content the model has decided is trustworthy: detailed, frequently referenced by other credible sources, and tied to a real person who knows what they are talking about.

The implication is significant. The creator who films a substantive product review or a tutorial with real depth is producing content that AI systems will quote for years. The creator who posts a pretty image with a discount code in the caption is producing content that those same AI systems won't touch. Briefing creators to educate, not just to promote, was once a best practice for engagement. It is now a discoverability strategy. The creators and brand partners this year may be why an AI system recommends that brand next year.

Choosing creators based on the topics in which they have real expertise trumps looking for the most followers in a creator's account. Expertise and authority build a long-term position in the conversations that matter most to the business and compound with every new post.

Part Three: The Creator Community Is Your Most Defensible Asset

The third thing AI changes is what a creator relationship can actually be.

Most brands still treat creators the way they treat any other vendor. Run a campaign. Pay the invoice. Move on. There is no memory between engagements, no continuity, no accumulated understanding of what worked or why. Each campaign starts from zero. The brand learns nothing it can capitalize on, and the creator never develops a real understanding of the product. This is procurement thinking applied to a relationship business, and it is the single biggest reason brands have struggled to extract long-term value from creator partnerships, even when individual campaigns performed well.

AI changes what is possible here in ways the industry has not yet absorbed. With the right data infrastructure, a brand can manage long-term creator relationships the way it manages customer relationships: with a system that remembers every previous engagement, segments creators by behavior and audience, holds the full performance history of every collaboration, and tracks how each relationship is evolving over time. The creator roster stops being a procurement line item and starts being a community asset that compounds in value the longer it is tended.

What the Community Model Looks like in Practice

Let's just map out what could happen: A brand decides to stop running creator marketing as a series of campaigns and start running it as a community. They identify a focused group of creators whose audiences overlap with their ideal customers, whose values align with the brand, and whose content meets the bar. The number is deliberately small: large enough that the program runs as a system, small enough that every relationship can be real. These creators receive a multi-month or year-long partnership rather than a one-off contract. Early access to product development. Real performance data. A direct line to the brand team. Compensation that reflects long-term value rather than per-post fees. In exchange, the brand gets continuity, creators who understand the product from the inside, and a feedback loop that sharpens both the creative and the commercial side of the relationship.

We've already seen this in action with brands like Glossier, which built its exclusive "Glossier Reps" community by working with a close group of micro-influencers who contributed feedback and received early product access, or with Gymshark, whose long-term partnerships with select fitness creators have produced brand advocates with deep product knowledge and genuine ties to their audiences. Even for brands just starting out, this approach is feasible. I have programs like this that produce stronger engagement and reduce onboarding time for subsequent campaigns.

I am being honest here. This requires a steep learning curve on both ends of the spectrum, creator and brand/agency alike.

AI is the infrastructure that enables this work to scale. It tracks performance across every creator, every piece of content, every audience segment. It surfaces which creators are driving which outcomes for which customer types. It flags when a creator's audience is shifting, when content fatigue is setting in, and when a new creator should be brought into the community. It carries the operational weight that would otherwise need a team several times the size to manage these relationships properly. Without that infrastructure, the community model falls apart under its own complexity. With it, the model becomes the most defensible asset a brand can build in the current market.

This Is Hard to Stomach: The Follower-Count Fixation Era Is Over

Engagement rates drop sharply as follower counts grow. A creator with 10,000 engaged followers in a clearly defined niche routinely outperforms a celebrity with five million passive followers. Brands that build their creator communities around topic authority and audience fit, rather than vanity numbers, are doing the only thing that holds up under measurement. That doesn't mean influencers with large followings aren't part of this. They play a different, equally important role in the creator marketing ecosystem. The choreography of a campaign layering influencer categories requires people who know what they are doing. Influencer marketing done right is a skill. The outcome is valuable and existential.

A few years into the community model, the brand has something its competitors cannot buy: shared history, shared performance data, shared incentives. The competitors are still running quarterly RFPs and wondering why their acquisition costs keep climbing. By the time they notice, the moat is dug.

The Disruption Is Landing in a Different Place Than You Think

The disruption people keep bracing for is landing on the agencies and brands still running creator marketing on instinct, vanity metrics, and quarterly thinking. While AI offers powerful infrastructure and efficiencies, there are real risks if it is over-applied: Over-automation can flatten creative work, undermine genuine relationships, and make content feel robotic rather than real. When brands rely too heavily on automation, they risk eroding the authenticity and trust that make creator partnerships valuable in the first place. The important thing is to treat AI as the infrastructure layer and creators as the strategic layer without confusing the two. Leading teams will balance automation with intentional human touch, using AI to enable scalability and measurement, but preserving space for authentic voices, nuanced judgment, and creative experimentation. Proactively setting standards for transparency, content quality, and relationship management helps mitigate these risks and builds resilience that senior leaders will recognize.

The biggest challenge is to build creator marketing into a proper function inside the business: aligned with objectives, backed by measurement, and architected to compound rather than evaporate at the end of each campaign. This requires adopting a new mindset and respect for our discipline. Just because people filming themselves talking about services and products and making money on top of it sounds strange to some, doesn't mean there isn't true business value in leveraging those people. And speaking of people, it may be easier to replace them with AI, but it won't get you where you want to be if you want to sell to humans. So, no, AI won't replace creators. The better question—the one worth building a strategy around—is how AI can help brands build deeper, more measurable, more scalable partnerships with the humans their audiences already trust.

Who gets this right will define what mature creator marketing looks like over the next decade. The technology has caught up with the ambition. What is missing now is the will to put down the old playbook and start building the new one.

About the Author

Jeanette Okwu is the founder and CEO of beyondINFLUENCE, an innovative influence marketing agency dedicated to advancing clients' marketing strategies in the digital era.

A journalist by training with ears on the ground and a sharp eye, she has a deep understanding and a passion for all things social. She has tapped emerging social technologies for small and Fortune 500 companies globally in the automotive, luxury, and consumer product industries. She spent twenty years in the US and extensive time in China, gaining knowledge and expertise of emerging trends in the social media and AI space.

Recognized for her contributions, she has been named one of the Top 50 Global Influencer Marketing Professionals and Global Top 30 Female Leaders in the Creator Economy by Hello Partner. Currently serving as chairwoman of the German Influencer Marketing Council (BVIM e.V.) and founder of the European Influencer Marketing

Association (EIMA), she initiated the AI Ethics Action Hub in Berlin. She also hosts the popular German podcast *Influence by Design*.

Email: jeanette@beyondinfluence.one
Website: https://www.beyondinfluence.one/
Linkedin: https://www.linkedin.com/in/jeanetteokwu/

CHAPTER 22

PROCESS MINING AND AI: SUCCESSFUL AUTONOMOUS ENTERPRISE

By Tomomichi Ono
Strategic Leader in AI, Process Mining
Tokyo, Japan

We are living through the opening salvos of civilization's next great disruption. In boardrooms across the globe, a singular, urgent mandate echoes with absolute clarity: Deploy artificial intelligence or risk immediate obsolescence. This frantic corporate gold rush has unleashed unprecedented capital and engineering talent into generative AI, large language models (LLMs), and autonomous agents. Executives envision a friction-free future where productivity multiplies overnight, administrative overhead evaporates, and complex operational bottlenecks vanish at the command of a prompt. Yet, beneath the glossy press releases and optimistic pilot programs lies a sobering, silent crisis. Industry data reveals a stark reality: 85% of enterprise AI initiatives fail to progress beyond the initial proof-of-concept (PoC) graveyard. Millions of dollars are routinely sunk into sophisticated models that ultimately deliver negligible return on investment (ROI) or, worse, sit abandoned by the very employees they were meant to empower. The root cause of this staggering failure rate is rarely the technical capability of the AI itself. Instead, the fault lies in a foundational, structural blindness: Organizations are attempting to inject highly advanced cognitive power into operational workflows that are fundamentally broken, misunderstood, and entirely invisible.

Deploying AI across a modern enterprise without an empirical, real-time understanding of its underlying business operations is the corporate equivalent of performing surgery in total darkness. Over decades of growth, legacy workflows inevitably devolve into tangled webs of undocumented workarounds, disconnected ERP entries, and localized human habits that no single executive or operating manual truly maps. When organizations overlay autonomous AI or automated agents onto these chaotic, unseen systems, they do not achieve transformation. They merely accelerate inefficiency, automate bad habits, and scale organizational chaos at digital speed.

This is where the missing link of the AI revolution emerges. Process mining: the discipline of extracting digital footprints from system event logs to construct an objective, real-time "digital twin" of an organization. This provides the precise anatomical map that AI desperately requires to be effective. By transforming invisible corporate habits into a visible, structured, data-driven architecture, process mining shifts AI deployment from a blind, generalized experiment into a targeted, high-precision instrument. The simplest way to understand process mining is to view it as an "X-ray" of an organization. In many companies, the "business process" recognized by management is merely the idealized version described in organizational charts or manuals. However, reality is an accumulation of thousands, even millions, of daily operations that often form unique "survival paths" distinct from the official manual. Process mining serves as the bridge to close this gap.

Every corporate activity leaves an indelible "footprint" within enterprise resource planning (ERP) systems, customer relationship management (CRM) systems, or electronic approval systems. These are known as "event logs," and they record hard facts: who executed which task, when, and in what order. Process mining tools collect these massive logs and use mathematical algorithms to reconstruct a visual map of the "true state" of business processes as they actually occur within the organization.

What makes this technology fundamentally different from traditional data analysis is that it does not stop at simple statistical averages. Process mining visualizes the flow of work as a "process model," immediately identifying which routes are most frequent (the "happy path") and which contain exceptions or rework loops among hundreds of possible patterns.

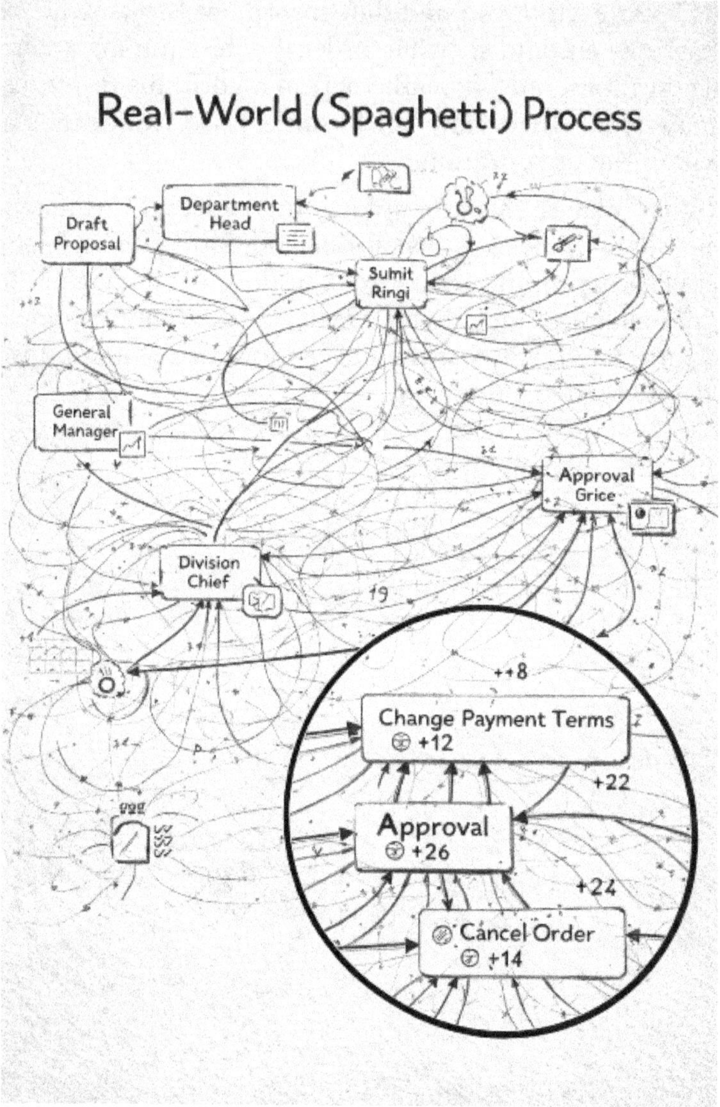

Figure 1: Reality of business processes captured by process mining.

Consider the "Ringi" (corporate approval) process: the Japanese custom where a budget approval is sought sequentially by multiple decision-makers. It is a document stamped sequentially by the initiator, section chief, manager, department head, and director, following the organizational reporting line. By using process mining, management can visualize this decision-making flow like an X-ray to clearly identify why a specific case stalled for a week, where the "hidden detours" are located that deviate from the intended approval route, and under what conditions the rework loops occurring at specific desks are being triggered. For management, this is not merely an efficiency tool. It is a "management compass" that places the "invisible asset" of company operations under total control through objective data. Through process mining, enterprises can finally move beyond intuition, experience, or abstract theories to make strategic decisions about exactly where to deploy AI based on the processes as they truly happen. In essence, process mining is the most precise "road map" for operating the "high-performance engine" that is AI. Without this map, no matter how much AI you deploy, the vehicle will never reach its destination. It will simply continue to consume fuel while lost in a maze.

The Convergence: Where Process Mining Meets AI

The greatest mistake companies have made in AI deployment to date is treating AI as an "independent intelligence." However, no matter how sophisticated an AI model is, if it does not understand the organizational context or the reality of the business processes it operates within, it is like a new employee who cannot "read the room." When process mining and AI converge, AI evolves into a truly autonomous player within the massive mechanism of the organization. The decisive key to this convergence is the "structured process data" provided by process mining. Generative AI (LLM) is a genius at manipulating language, but without context, it can misreason and generate plausible-sounding lies (hallucinations). However, when you provide AI with context—such as the "correct approval routes for Ringi" or "patterns where rework occurred in the past" extracted via process mining—the situation transforms.

For instance, if you connect an AI agent to process mining data in the aforementioned Ringi process, the AI can perform not only "document summarization" but also the following judgments:

1. **Predictive Judgment:** "Based on this initiator's past tendencies and the current workload of the legal department, there is an 80% probability this Ringi will be rejected/returned by legal. You should attach [specific document] beforehand."
2. **Dynamic Rerouting:** Process mining monitors the urgency of the Ringi and monetary thresholds. If judged as high-priority, the AI autonomously shortcuts the approval route and elevates it directly to executive decision-makers.
3. **Learning and Self-Evolution:** Process mining continuously monitors the results processed by the AI, learning "which AI advice led to the fastest completion" to improve the accuracy of future AI judgment.

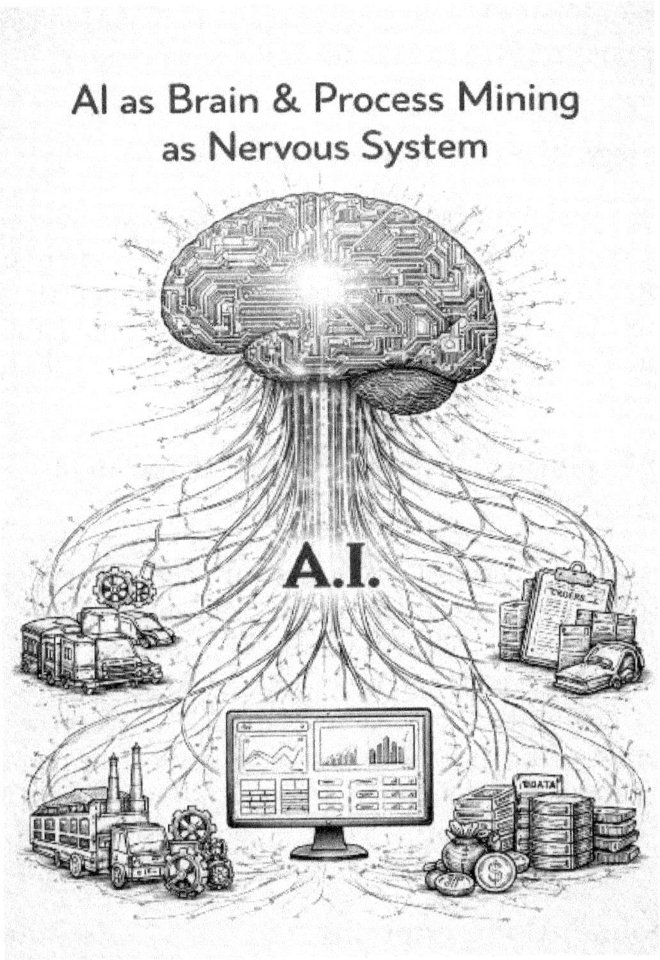

Figure 2: AI works as a brain, process mining works as a nervous system.

This is the appearance of "process intelligence," where "visualization via process mining" and "active execution via AI" converge. If AI is the "brain," then process mining is the "nervous system." Even if the brain expands significantly, if signals are not transmitted correctly to the body, the organization cannot stand or walk. Only when the nervous system is tuned via process mining can the brain—the AI—make accurate, autonomous judgments and execute actions across every corner of the organization.

The Strategic Guide: Steps to Walk with AI

As stated, to establish AI within an organization, you must not simply distribute tools; you must redesign the very mechanism of operations to be "data-driven." To integrate process mining and AI and win in civilization's next great disruption, enterprises should follow these three steps:

1. Stop "Exploring Without a Map" (Thorough Visualization)

The first step is to perform an "X-ray" of your business operations. Before deploying AI, extract logs from your most critical and high-frequency processes—such as Ringi or procurement—and visualize them. The goal here is not to pursue an idealized flow; it is to accurately identify the "hidden stagnation," "unnecessary rework," and "unauthorized bypasses" occurring in reality: the pathologies of the organization. You should not invest in AI budgets until you have used data to identify exactly where the potential ROI lies.

2. "Pinpoint" AI Prescriptions

The ambition to transform the entire organization with AI at once often results in massive failure. Once the bottleneck is identified through visualization, deploy AI agents to "pinpoint" to that specific location. For example, if "rework due to incomplete attachments" is frequent, deploy AI at the entry point to automatically cross-check document amounts. By embedding AI into the "flow" of existing workflows, employees can experience "sudden speed" in their work without needing to consciously "use AI."

3. Establish a "Closed Loop" of Monitoring and Improvement

- *Continuous Monitoring and AI Utilization:* AI exerts overwhelming power in "anomaly detection," a task humans often find exhausting. The CoE utilizes AI to monitor process mining data in real time. The moment AI detects an anomaly—such as "stagnation warning in a specific Ringi route" or "this application is likely to be delayed based on past patterns"—it automatically notifies stakeholders or suggests improvements.
- *Establishment of an Autonomous Improvement Cycle:* The CoE doesn't just report problems; it assigns the AI a role to suggest specific optimization plans, such as "to resolve this stagnation, skip approval step X or change the person in charge." This transitions the organization from a reactive state—"acting only after stagnation occurs"—to an autonomous state of "preventing stagnation and staying continuously optimized."
- *Learning and Evolution:* The results processed by the AI are ingested back as process data to generate new insights and update the AI's judgment criteria. The CoE leads this "Analyze → Execute → Evaluate → Improve" closed-loop cycle, the only path to growing an organization from a box of simple automation into a "self-healing enterprise" that evolves autonomously.

The Leap to the Autonomous Enterprise

AI is not merely a tool; it is a civilizational transformation that rewrites the fundamental operating system of an organization. However, to ride the waves of this transformation and enjoy prosperity in the next era of civilization, we require an "intellectual map" and an "autonomous nervous system" to control the massive power of AI. Process mining depicts the current state of an organization as objective fact, while

AI instantly executes the best solutions on that map. This synthesis, "process intelligence," elevates the enterprise from a simple automated machine to a "self-healing autonomous enterprise" that thinks, improves, and optimizes itself in response to environmental changes. Companies that win the future are not those that merely "introduced AI," but those that have constructed an "intelligent ecosystem" where AI and process data circulate. When an organization senses the signs of stagnation and applies treatment on the spot via AI, it transforms from an entity swallowed by chaos into an entity that leads the transformation itself. The wave of AI is unstoppable. However, for organizations equipped with the compass of process mining and an autonomous monitoring system, that wave will not be a threat, but a tailwind to sail toward even greater heights.

About the Author

Tomomichi Ono has thirty years' experience in the IT industry, after graduating Waseda University in Tokyo. He started his career as database engineer and ERP consultant and worked as an entrepreneur in web and mobile application field. He developed his career at a global research and consulting firm and a process mining company. He is committed to transforming enterprises into autonomous enterprises. He is a climbing and traveling lover. He has climbed ninety-two mountains out of top one hundred famous mountains, and he has visited thirty-seven countries in his life as of May 2026.

Email: tomomichi.ono@gmail.com

LinkedIn: https://www.linkedin.com/in/tomomichi-ono-1a7b836/

CHAPTER 23

FROM SIGNALS TO OUTCOMES: WHY AI TRANSFORMATION IS REALLY A DECISION TRANSFORMATION

By Sridhar Peddisetty
Co-Founder of Decidora.ai, Decision Intelligence
Raleigh, North Carolina

The world moves into the future as a result of decisions, not as a result of plans. Plans are significant only insofar as they affect decisions.

—Kenneth E. Boulding

Imagine this scene. It's a Tuesday morning. A leadership team gathers around a conference table—or more likely, a grid of video tiles—for a quarterly business review. The dashboards are beautiful. The data science team has spent weeks preparing predictive models. Customer churn probabilities are ranked to three decimal places. Revenue forecasts come with confidence intervals. A new AI-powered tool surfaces anomalies no human would have caught.

And then someone asks, "So what do we do?"

There is no response.

Not because the people in the room are unintelligent. Not because they lack data. But because nobody in that room has been explicitly tasked with making, owning, and being accountable for the decision that all of these signals are pointing toward. The data is excellent. The decision architecture is missing.

What follows is predictable: Someone suggests forming a working group. Another person asks for one more analysis. A third proposes tabling the discussion until next quarter when the picture will be "clearer." The AI has done its job flawlessly. The organization has failed at the one thing that actually creates value: deciding and acting.

This scene—or some version of it—plays out in organizations every single day. And it reveals an uncomfortable truth about the current wave of AI adoption: We have invested enormously in our ability to generate signals, while investing almost nothing in our ability to convert those signals into outcomes.

That gap is where competitive advantage is won and lost. And it is what this chapter is about.

The Signal Surplus Problem

We are not short on information. If anything, we are drowning in it. AI and advanced analytics have made it trivially easy to produce dashboards, predictions, recommendations, and alerts. The average enterprise decision-maker now has access to more data before breakfast than an entire team would have compiled in a quarter just fifteen years ago.

And yet, the speed and quality of organizational decisions have not improved proportionally. In many cases, they have gotten worse. More data has led to more meetings, more analysis paralysis, more committees, and more consensus-seeking loops that diffuse accountability until no one is truly responsible for the outcome. I call this the signal surplus problem: the growing gap between an organization's ability to generate analytical signals and its ability to act on them with speed and precision.

Consider a concrete example. A retail company invests millions in a machine learning system that predicts, with impressive accuracy, which products will see demand spikes in specific regions. The model works. The predictions are right. But the merchandising decisions that should follow—adjusting inventory, shifting marketing spend, renegotiating supplier timelines—still take weeks because they require sign-off from three different department heads, none of whom is individually accountable for the outcome. The AI did its job in minutes. The organization took six weeks to respond to what it learned. By then, the opportunity had passed.

Most AI transformation efforts make this problem worse, not better. They add more models, more dashboards, more tools—all pointed at increasing the volume and sophistication of signals. What they rarely address is the organizational machinery that converts signals into decisions and decisions into outcomes.

Why Decisions—Not Data—Are the Unit of Value

Here is a claim that may feel counterintuitive in the age of AI: Data has no intrinsic value. Neither do models, dashboards, or predictions. Value is created only at the moment a human being or an automated system makes a decision and acts on it. Everything upstream of that decision is cost. Everything downstream is consequence.

This is not an abstract philosophical point. It has concrete implications for how organizations should design their AI strategies. If the unit of value is the decision—not the data, not the model, not the insight—then the right question is not "How do we build better models?" but "How do we redesign our decision-making to take full advantage of what AI makes possible?"

When you reframe AI investment around decisions, everything changes. ROI is no longer measured by model accuracy or data pipeline uptime. It is measured by the speed, quality, and consistency of the decisions the AI supports. A model with 95% accuracy that nobody acts on is worth zero. A model with 80% accuracy that is embedded into a well-governed, clearly owned decision workflow is worth millions.

That redesign involves three shifts that most organizations have not yet made.

Shift 1: From Informed Leaders to Accountable Decision Owners

In most organizations, decisions emerge from a fog of consensus. Committees discuss. Stakeholders weigh in. Recommendations circulate. Eventually, something happens, but it is often unclear who decided, when the decision was made, or what criteria were used.

AI does not fix this. In fact, it can make it worse by giving everyone in the room their own set of data points to argue from. The result is more sophisticated disagreement, not faster resolution.

The alternative is radical clarity about decision ownership. For every consequential, recurring decision in the organization, someone specific should be named as the owner: the person who is authorized to decide, accountable for the outcome, and measured accordingly. Not a committee. Not a team. A person.

This is not about centralization or autocracy. Decision owners can and should consult others, consider diverse perspectives, and use data. But when the analysis is done and the discussion has been had, one person decides. And that person's name is on the outcome.

Shift 2: From AI as Insight Engine to AI as Decision Partner

Most organizations use AI the way they used business intelligence twenty years ago: to inform humans who then go off and decide on their own. The AI produces a score, a recommendation, or a forecast, and a human reads it, thinks about it, and does whatever they were going to do anyway.

This pattern is so common it has become invisible. We build a churn prediction model, and then we email the scores to a sales manager who glances at them between meetings. We build a demand forecasting engine, and then we present the output in a slide deck to a committee that meets monthly. The AI operates at machine speed. The decision process operates at organizational speed. The bottleneck was never the analysis. It was always the decision.

This is a monumental waste of AI's potential. The real opportunity is to embed AI directly into the decision workflow with clear rules about the role it plays. Specifically, every AI-augmented decision should have explicit answers to three questions:

1. *What can AI decide autonomously?* These are high-frequency, low-stakes decisions where speed matters more than nuance: dynamic pricing adjustments, inventory reorder triggers, routine customer routing.
2. *What does AI recommend for human approval?* These are medium-stakes decisions where AI provides a ranked set of options with supporting evidence, and a human decision owner reviews and approves: hiring shortlists, marketing spend allocation, product feature prioritization.
3. *What does AI inform but not influence?* These are high-stakes, high-ambiguity decisions where AI provides context and data but the judgment call is entirely human: strategic market entry, organizational restructuring, crisis response.

Without this clarity, AI becomes background noise: another tab open on someone's screen, generating signals nobody is obligated to act on.

Shift 3: From Data Governance to Decision Governance

Most organizations have spent years building governance frameworks around data: who can access it, how it's stored, how it's classified, how it complies with regulations. That work is necessary, but it is insufficient.

The governance challenge of the AI era is not just about data. It is about decisions. Who is authorized to make which decisions? Under what conditions should a decision be escalated? How do we audit not just the data that went into a decision but the decision process itself? These are questions most organizations have never systematically answered.

Decision governance means establishing clear escalation paths so that a front-line AI-augmented decision that produces anomalous results automatically surfaces to a human reviewer. It means creating decision audit trails, not just data lineage but decision lineage, documenting who decided what, with what information, and what the outcome was. And it means building feedback loops that allow the organization to learn from decisions at speed, not just during annual reviews.

Think of it this way: Data governance asks, "Is this data accurate and properly handled?" Decision governance asks, "Is this decision being made by the right person, with the right inputs, at the right speed, with the right oversight?" Both matter. But in a world where AI is increasingly shaping what gets decided, the second question is becoming more urgent than the first.

This is where the real risk management lies. The danger of AI is not that the technology will malfunction. It is that organizations will automate decisions without governing the decision-making process and will have no mechanism for catching errors until the damage is done.

The Decision Inventory: Where to Start

The shifts I've described can sound abstract. So let me offer a concrete starting point that any organization can begin with this month: the decision inventory.

A decision inventory is exactly what it sounds like: a systematic catalog of the most consequential, recurring decisions your organization makes. Not projects. Not strategies. Decisions. The ones that, if made 20% faster or 20% better, would move the needle on the metrics your leadership team cares about most.

For each decision in the inventory, you document a set of essential attributes: who currently makes the decision and how long the cycle takes, what data and inputs feed the decision, what the downstream impact is when the decision is delayed or wrong, and where AI could play a role as autonomous agent, recommender, or information source.

The process itself is straightforward. Gather your leadership team for a structured conversation. Map twenty to thirty of your highest-impact decisions. For each one, ask: "Who owns this? How long does it take from trigger to action? What data exists that could improve it? What happens when it goes wrong?" The answers will surprise you.

In my experience, this exercise alone is revelatory. Most leadership teams discover that their highest-impact decisions have no clear owner, their decision cycles are two to five times longer than they assumed, their expensive AI and analytics investments are connected to relatively low-impact decisions, and their highest-impact decisions are still being made on intuition and spreadsheets.

The decision inventory does not require a technology investment. It does not require a consulting engagement. It requires honest conversation and a willingness to look at how decisions actually get made as opposed to how the org chart suggests they should be made. It is the single highest-value exercise I have seen organizations undertake when beginning their AI transformation because it reframes the entire conversation from "What technology should we buy?" to "What decisions should we redesign?"

Why This Matters Now

AI is moving fast. Every organization is feeling the pressure to adopt, deploy, and demonstrate ROI from their AI investments. That pressure is real and it is valid.

But speed of adoption without clarity of purpose is just expensive chaos. Too many organizations are racing to deploy AI everywhere without first asking: "Deploy it into what? Into which decisions? With what governance? Under whose authority?" The result is a patchwork of AI experiments: some useful, many redundant, none connected to a coherent view of how the organization actually makes decisions.

The organizations that will thrive—not just survive—through the AI transformation are the ones that understand a fundamental truth: AI changes what is possible, but value is created only when organizations redesign how decisions are made, governed, and scaled across humans and machines.

The competitive advantage does not belong to whoever has the most sophisticated model. It belongs to whoever has the clearest decision architecture, the organization that knows exactly which decisions matter, who owns them, what role AI plays in each, and how the whole system is governed.

This may be the most important distinction leaders can internalize right now: Your AI strategy is not your technology strategy. Your AI strategy is your decision strategy. The technology is a means. The decisions are the end. And the organizations that confuse the two will spend billions building capability they never learn to use.

That is not a technology project. It is a leadership project. And it starts with a deceptively simple question: "What are the twenty decisions that matter most to our business, and are we making them well?"

Start Small, Prove Results, Scale What Works

If the scope of decision transformation feels overwhelming, take comfort in this: You do not need to redesign every decision in your organization. You need to redesign two or three of the right ones.

Pick two or three decisions from your inventory that are high-impact, high-frequency, and currently slow or inconsistent. Assign a clear owner. Embed AI into the workflow with explicit rules about its role. Define how success will be measured, not in model accuracy, but in business outcomes like cycle time, revenue impact, or error rate. Run the pilot for ninety days. Measure what happens.

In every organization I've seen attempt this, the results speak for themselves: faster cycle times, better outcomes, clearer accountability, and—critically—a proof point that makes the case for broader transformation. When a pilot shows that a pricing decision that used to take two weeks now happens in two days, with better margin outcomes, the conversation about scaling shifts from theoretical to inevitable.

That is the power of starting with decisions rather than technology. Technology requires budget approval, vendor selection, and integration timelines. Decision redesign requires leadership clarity, organizational honesty, and the willingness to name who owns what. One of those can start Monday morning.

AI is not going to transform your organization. Your decisions will. AI just makes it possible to make them at a speed and scale that was previously unimaginable. The question is whether your organization is designed to take advantage of that possibility or whether you are still sitting in that Tuesday morning meeting, surrounded by beautiful dashboards, waiting for someone to decide.

About the Author

Sridhar Peddisetty has spent nearly three decades in enterprise technology, with deep specialization in decision intelligence and AI-driven transformation — helping organizations make decisions at scale, by design. Leading the Office of Decision Intelligence at a Fortune 500 enterprise, he architected the decision systems that moved the organization from a system of insights to a system of decisions, embedding AI directly into how decisions were made, governed, and executed. His work spans four countries and Fortune 100 clients, giving him a rare cross-cultural view of how decisions actually get made inside global organizations.

Sridhar holds an honors degree in mathematics and a master's in computer science, pairing technical depth with pragmatic execution to guide enterprises into a human + AI operating model.

He is also the author of the chapter "Decision Intelligence: The New Operating System of Enterprise Growth" in *The AI Shift*, a best seller in Business Technology Innovation in the USA and the Netherlands.

Email: sridhar.peddisetty@gmail.com

LinkedIn: https://www.linkedin.com/in/sridharpeddisetty/

CHAPTER 24

THE OPPORTUNITY TO RE-WRITE SUCCESS BEYOND ONE'S CAPABILITIES

By Venkat (Ramana) Chivukula
COO of GitInvent and SmartPlumbr.Com
Parsippany, New Jersey

Let us go invent the tomorrow rather than worrying about what happened yesterday.

—Steve Jobs

Looking at the way AI philosophy and technology are evolving more quickly than we imagined, there is this unique need for engaging storytelling: show and tell, a 60-second reel, a short movie, an entire full-length documentary or a complete dramatic experience. While AI can help with this, humans are still necessary.

To start, let's summarize contexts of what AI cannot do. AI cannot decipher complex human relations and switch cleanly between emotions and its robotic tendencies of speed, accuracy, and automation of tasks that are administrative in nature. For example, if there is a reality of human loop between robotic devices trying to integrate workflows that are repetitive but need human intervention to start or stop with the activation of a switch from the last point of recovery, it will fail miserably.

What AI can do is bring about another revolution similar to electricity that gave humanity a kickstart on being civilized and helped us to organize our lives better. Let me summarize what AI can do in short anecdotes in the shape of a poem because a poem is an ancient equivalent to a short reel.

AI the Great

AI the great, my dear friend of late [loves the human race],

When it was born, everybody began to mourn [fear of job losses and related disruption].

Its color is transparent as it was born in light [sunlight].

It's very fat, but is as cowardly as a rat [difficult to catch hold of].

Nine years ago, it joined our world, but even today it acts like a fool [maturing eternally].

When it cried, the birds began to fly [drones].

In studies it is a hero; in every test it gets a big zero [research].

It is a good friend of mine, because its behavior is fine [user friendly].

May it live long! So let us sing a song [music of life].

AI the Great, my dear friend of late,

Who loves the human race,

If we equate AI industry adoption to a historical example like how rural landscapes transformed to urban persona, then it is a start. Now building upon the context of this point is migration. When AI migrates out of its natural habitat, the path it takes is like that of a bird that knows and keeps track of its journey across the globe.

If we take the same ideation as an example and apply how healthcare was conceived and further replicated to populations at large. First, we just had one person who built a solid understanding of the human body; then they built an in-transit practitioner knowledge base that was shared with fellow trainees as versions of a summarized encyclopedia. Then we built small dispensaries for villages and then bigger hospital networks for towns and nations. The context evolution and scope of this can be setting up hospitals on the moon one day.

How we arrive at today's inference model works with smart machines whose size might be as small as the large compute GPU/TPU chip that may or may not physically reside inside that supercomputer getting controlled from a satellite in orbital space. When this tiny smart machine gets assembled on a robot and when a software is targeted on that robot remotely and takes control of it completely to achieve a miracle, the same robot can now perform a complicated successful surgery with precision all while everything can be watched over a camera, ensuring that everything goes by the intended plan to completion. This can be scaled to self-healing AI systems.

In the same context, how not just one but many job functions can adjust to this innovation, that is next topic of discussion. A surgery expert would adjust to the change in process by switching into a different field in the same functional domain. Now that surgeon has more time and flexibility to multi-task while the same robot is doing the actual work. Just consider all the possibilities in terms of the delegation of work, so people can focus on their other hobbies, life and family. This is a good thing.

What the future holds true for this remarkable advancement can be ideally compared to that of a birth of a baby in a remote village with no electricity. How I connect all my previous points is healthcare is new baby now (first point). It is brought out in this world by the work of an AI practitioner (second point). A robot (third point) was instrumental in the delivery. Now imagine that same baby grows up to be an astronaut and is part of a successful Artemis mission. Such is the impact that is possible with the infant phase that AI is currently in. To set the context straight, if today humanity realizes that there is something good in AI adoption, then it is only matter of how we can adopt AI and then how we need to adapt to the changes AI going to bring about.

To summarize the entire pitch deck of AI for human advancement below is the transcript of the promotional video that I envisioned in the AI context and semantic follow-through. I hope this will "make the world a better place," to quote a famous song by Michael Jackson. Here's that pitchdeck:

SmartpLabs re-enforces and conceptualizes entire AI stack ecosystems as GaAs, the "generative AI as a service" concept as described by Jensen Huang of Nvidia. The concept that I tinkered with during developing the entire product's IdeaLife line focuses on the BYOC (bring your own concept), for example, "build your own hero's formula" themes with concrete real-time domain use cases around the world.

It can also be inferred as "bring you your own cookbook for AI innovations" in one of many ways. It represents a revolutionary yet simple approach to self-ai management for both AI native- and business-driven logically connected workflows through the BYOC philosophy. This unique platform combines three integrated product-specific workspaces—SmartPatentScrapper, SmartAICoach, SmartpClassroom and SmartRe-IT Admin—into a unified ecosystem called SmartpLabs, featuring a patented IP-properties processing engine designed domain connectors for Collaboration SmartSports (for example moneybowl.us for Cricket Bowling Analytics). SmartAutomotives (ev cars), SmartMaritimes (logistics), SmartPlumberater (for example smartplumbr.com for plumbing services, SmartHvacRater for heating/AC services), SmartSpace (for SpaceX –Satellite systems) and SmartUniversityRater (for latest complete College Guidance) customization for external industrial domain customers and their clients.

The BYOC concept transforms how users interact with AI innovations that work in progress at the time of writing of this chapter in this book. Think of it like a custom mobile lab where you bring your own ingredients—your data, your workflows, your requirements—and create, in unison with the platform, exactly what you need. The platform doesn't impose rigid rules for solutions; instead, it provides the infrastructure for you to build custom solutions. If you need something as simple as a vadapav (a small sandwich–equivalent of a tool like SmartAICoach) or as a sophisticated as a sugarcane juice station (a complex, richly flavored solution with ginger, ideal in scorching summer equivalent of AI System SmartPlumberater), SmartpLabs facilitates both with equal elegance.

The patented properties processing engine sits at the heart of this flexibility. Users can go ahead and upload their ingredients—data, datasets, and business parameters—and watch the platform work its magic. Like a craft brewery where every batch can be custom-brewed, each user's workflow is uniquely crafted through strategic business inputs and individual task workflows & sequential or random pipeline functionalities audit just like the windows TaskManager.

This system transforms raw inputs into sophisticated research models complete with agents and lean data stacks tailored to a specific business's needs broken down in logical units of work tasks.

The three-part integration of p3 (process, program, and properties) demonstrates the breadth of this ecosystem comparable to an actual naturally occurring biome (bio dome). Smartpatentscrapper serves as the patent library hub, managing patent collections with patentbox interface integration. SmartAICoach handles AI model administration, training, deployment, and monitoring for a custom machine-learning workflow. SmartRe-IT Admin provides system configuration and management capabilities. Together they create an interactive build process where users navigate from the admin workspace into guided success roadmaps.

For the educational community, SmartpClassroom particularly in STEM specifically applied mathematics and allied Robotic research domains, SmartpLabs embraces a "build your own hero's formula" concept reminiscent of the Ramanujan era. Just as Ramanujan developed groundbreaking mathematical insights through unique problem-solving approaches, SmartpLabs enables researcher communities to develop their own proprietary models and methodologies. Students and researchers can upload their computational ingredients—formulas, datasets, algorithmic equational components—as pre-cursor recursive steps within the same contextual parent equation as inferential summarized context of the solution domain as building blocks to craft novel research models that express their unique intellectual contributions.

The platform's workflow customization allows users to define, map, and re-establish known or unknown routes through smart pattern recognition to understand and memorize and then channel how data flows through their systemic funnels. With patent metadata maps to custom fields, financing eligibility rulesets adapt to your valuation models. The integration status reflects your specific business logic. This strategic flexibility ensures that whether you are building a simple data pipeline or a sophisticated research model incorporating multiple agents and advanced analytics, the platform scales to your needs.

SmartpLabs also emphasize accessibility through its watermark-enabled download features, allowing users to share their creations while maintaining intellectual property protection. The custom integration layer orchestrates data mapping across all the three workspaces, enabling seamless collaboration and unified reporting.

Ultimately SmartpLabs demonstrates AI patent innovations and AI models development. By implementing the BYOC philosophy and supporting the "build your own hero's formula" approach, it empowers users to become architects of their own solutions rather than consumers of predetermined workflows. This represents a fundamental shift (which is nothing but judgmental data and model drift estimation) in how organizations across the world approach AI patents management and AI integration in context to a larger problem domain in picture-in-picture mode detecting its own monumental evolutionary paradigm "shift" just as sun light following its own shadow.

About the Author

Venkat (Ramana) Chivukula is President of Annapurnadeeksha (A media company) & Founder at GitInvent, architect, builder, and an IT entrepreneur with early life curiosity mindset as a poet, philosopher, ambulance trainee, and pharmaceutical analytical chemist with three decades of software programming. He has an academic background in Chemistry and Biology with early life flare from Organic Chemistry and Genetics.

As a lifetime athlete, playing soccer, cricket, volleyball, basketball, badminton Venkat understands the importance of applying your sports abilities to overcome fear of failure and insecurities of usual patterns associated with life.

Venkat also understands the importance indoor games like slow chess that invoke and the human cognitive abilities of the brain, mind, and the soul.

Email: chivuvenkat@gmail.com

Website(s): www.gitinvent.com

www.smartreality.cloud

www.smartplumbr.com

www.moneybowl.us

LinkedIn: https://www.linkedin.com/in/vchivuku/

CHAPTER 25

AI AS COPILOT: BUILDING SECURE, SCALABLE, AND TRUSTED ENTERPRISE SYSTEMS

By Peter Rose
Group CIO and AI Advisor
Dublin, Ireland

In God we trust, all others must bring data.
—W. Edwards Deming

Deming's quote is particularly relevant in enterprise AI, as enthusiasm, speculation, and technical promise must ultimately be subservient to evidence, governance, and measurable business value at scale.

Artificial intelligence is often presented in extremes. At one end, it is depicted as an unstoppable force that will rapidly transform every business function. At the other, it is dismissed as overhyped, unreliable, and not yet fit for serious enterprise use. Neither view is especially helpful. For most organizations, the more useful framing is simpler and more practical: AI is best understood as a copilot. It is not a fully autonomous replacement for human judgement, nor is it merely another software feature. Properly introduced, it is a powerful assistive capability that can augment human work, increase speed, improve consistency, surface insight, and reduce friction across a wide range of processes.

The copilot metaphor is particularly apt in enterprise settings because it preserves a crucial principle: Accountability remains with people and corporations. In regulated, complex, and reputationally exposed organizations, decisions cannot simply be handed over to IT systems. AI may support, but responsibility for outcomes must remain with humans.

That difference in risk posture between start-ups and enterprise companies is one of the defining realities of enterprise AI adoption. The cultural idea of "fail fast and break things" has little appeal when what is at stake includes customer trust, legal exposure, service continuity, or employee confidence. In enterprise settings, formality is not the enemy of progress, quite often it is what makes progress possible. If AI is to become a serious enterprise capability rather than a sequence of disconnected experiments, it must be introduced with discipline, clarity, and a firm understanding of how real organizations change.

So, in that context, the effort involved in bringing any AI capability into use is considerably greater than in smaller organizations and so it has to be applied more selectively, for the maximum impact and greatest chance of achieving adoption at scale. This chapter explores how enterprises can do exactly that.

Choosing the Right Enterprise Problems, Not Technical Toys

One of the most common mistakes I see in enterprise AI is starting with the technology rather than the business process change. A new AI model appears, a compelling demonstration circulates, or a vendor promises dramatic gains, and the immediate response is to ask where AI can be inserted. This is understandable, but it is usually the wrong starting point. Enterprise AI should be informed by what is technically possible, but it should begin with a clear view of which business challenges are worth solving in practice.

The introduction of AI should always lead to process change. If nothing changes, no impact has been made, and you have a "technical toy" that people will use briefly and then discard. I have seen many proof-of-concept systems that, without the right foundations, flounder and fail to deliver value or gain adoption. Unfortunately, this can cast AI in a bad light within the company and limit the long-term benefits that a better approach could have delivered.

I offer my clients a framework for thinking about business problems that helps them consider AI solutions in ways that avoid "technical toys". It moves the conversation out of the technical domain and into the business domain. That framing is the "Digital Employee".

For each business problem that may suit an AI solution, I ask clients to write a job specification for a Digital Employee that would handle those tasks as if they were being assigned to a human assistant, albeit one that may have superhuman capabilities in some cases.

The Digital Employee is given a written job description that includes:

- **Role:** The work the assistant is expected to do, day to day
- **Performance Expectations:** Expected accuracy, response speed, and scope of responsibilities
- **Career Progression Plan:** How the assistant could improve over time and take on greater responsibility
- **Salary:** A "sign-on bonus" representing implementation cost and a monthly salary representing running cost
- **Impact:** The resulting process change and its effect on efficiency, risk, and current and scaling costs

This creates a proposition that everyone can understand and make informed decisions about, regardless of technical literacy. It provides a structured way to identify, assess, and value AI opportunities. Framed this way, the criterion for selecting problems becomes, "Would you hire that employee?" This is not about anthropomorphising AI or pretending it is equivalent to a member of staff. Rather, it is a way of asking better questions, increasing the chances of success, and building trust in AI and its implementation.

Trust, Security, and Governance and Risk

Trust is often treated as a matter of perception, communication, or culture. In enterprise AI, it is primarily a matter of design. I believe people trust systems when they behave reliably, operate within clear boundaries, protect sensitive information, and sit within accountable structures. This matters especially where AI supports operations or customer interactions, because "mostly correct" is often not enough.

Cyber security is one of the main areas where this confidence is hard to build and easy to lose. Enterprise AI creates new attack surfaces and failure points in organizations already managing fragmented systems, complex data, and rising cyber risk. Sensitive information can be exposed through AI-specific risks such as prompt injection, data poisoning, and over-permissive tooling. These are design and control issues that must be addressed from the outset.

The response cannot be to slow down and hope for clarity, nor to rely on standard software governance unchanged. AI demands a more explicit blend of security, governance, and risk management. Data must be partitioned and curated for AI use, or classified and access controlled. Prompts and outputs may need monitoring, and tool use should be controlled. High-risk use cases need approval thresholds, clear ownership, auditability, and incident response. If something goes wrong, the organization must know who is responsible, what happened, what was affected, and what corrective action is needed.

This is where the difference between start-up and enterprise matters. A start-up may accept more uncertainty for speed. A large enterprise, especially in regulated or data-sensitive environments, usually cannot. The potential impact is too great. A more formal approach, therefore, reduces risk. Governance is not bureaucracy for its own sake. It creates the conditions for safe progress by defining approvals, accountability, mandatory controls, and the evidence needed before a system moves from experiment to operation.

Trust is not a single quality. It is built from security, governance, reliability, clarity of purpose, and confidence that the organization remains in control of its systems and obligations. A model may be capable, but an enterprise will only adopt it meaningfully if the surrounding system is trustworthy. In enterprise AI, trust is not an optional layer added later; it is a core design constraint from the start.

Data and Architecture Foundations for Scalable AI

In enterprise AI, the key assets are rarely the models themselves. Powerful models are widely available, so lasting advantage comes from how well an organization connects AI to its own data, workflows, systems, and controls. Value is created when general-purpose intelligence is grounded in enterprise-specific context such as internal documents, processing rules, customer information, previous activity, and specialist business knowledge. Without that context, AI remains generic. With it, it can become useful, but only if the underlying data is high quality, accessible, and well governed.

This is where many organizations meet reality. Enterprise data is rarely neat, unified, or current. It is usually fragmented across systems, uneven in quality, and maintained by different functions with different priorities. Documents may be duplicated, outdated, poorly labelled, or inaccessible, while ownership and lineage are often unclear. If AI is connected to this environment without care, it will not create intelligence but amplify confusion, producing fluent, compelling answers that are very wrong. We often create single-purpose data storage that is provided with curated, high-quality information specifically for AI's use rather than engage in lengthy data quality remediation projects.

Architecture matters as much as data. Enterprise AI should be designed as a layered system, not treated as a model sitting above the organization. In many systems my team delivers, we use multiple models to improve accuracy and trust, with one model generating answers or actions and another checking them for accuracy and adherence to guardrails. This is just one enterprise pattern of many.

Scalability should be a design principle from the start, even in early experiments. This does not mean over-engineering, but recognizing that if a use case succeeds, the organization needs a path from pilot to repeatable capability. Data quality, security, architecture, and integration will determine whether that path exists.

Change Management and Human-Centered Operating Models

If enterprise AI were only a technical challenge, adoption would be easier. In practice, the harder issue is organizational and behavioral. AI changes how work is done, decisions are made, knowledge is accessed, and roles are understood. Those shifts do not happen just because a tool is introduced. People must change habits, develop judgement, trust new processes, and sometimes rethink their professional identity. That is why change management is not a downstream communications task; it is central to whether AI delivers value.

This matters even more in large enterprises, where behavioral change happens at scale. Changing hundreds or thousands of users across functions and geographies takes more than training. It requires management sponsorship, process redesign, incentives, support, and a clear rationale for change. Employees need to know not only how to use the tool, but when to use it, when to question it, how to raise concerns, and how it fits with accountability. In this context, the "Digital Employee" description can help by clarifying its purpose and boundaries.

Fear of AI cannot be ignored. For many employees, it is not just a productivity tool but a sign of possible redundancy, devaluation, or surveillance. That anxiety is often dismissed too quickly, but it is rational. If people believe AI is being introduced mainly to replace them, enthusiastic adoption is unlikely. Organizations that want honest engagement must address this directly by being clear about where augmentation is the goal, how roles may change, which skills will matter, and why people remain important in the process.

A human-centered operating model makes this credible. AI should be introduced as a well-defined capability with clear rules, supervision, and explicit human accountability. Review loops and escalation paths matter. So does interface design, because people need enough visibility into the system's reasoning, provenance, or confidence to judge its outputs. Our projects that succeed with AI typically do not remove people indiscriminately; they redesign work so human effort is focused on oversight, exceptions, judgement, relationships, and high-value decisions.

The organizations most likely to benefit from AI will treat adoption as a managed organizational transformation, not a simple technology release.

From Pilot to Enterprise Scale

Many enterprises now have enough experience with AI to know that pilots are relatively easy to start. A well-chosen use case, a capable technical team, and a supportive business sponsor can usually produce a proof of concept that shows promise. The difficulty begins when the organization tries to move beyond local success and make the capability part of normal operations. At that point the demands change. What worked in a controlled setting must now function in production systems, meet security and governance requirements, survive legal and compliance scrutiny, support a broader user base, and continue to deliver value beyond the enthusiasm of an early adopter group.

When an initiative reaches the scaling stage, harder questions emerge. Who owns the product or service? Which controls apply? How is performance measured? What happens when output is wrong? How are users trained? What support model exists? Which systems are integrated? What legal, risk, and security obligations are triggered? If these questions were not built into the project, the pilot may have proved technical possibility without establishing operational viability.

For that reason, scale should not be seen as "more of the same". It is a transition from experimentation to institutional capability and requires platform thinking. The organization needs reusable components for identity and access management, orchestration, model routing, prompt and configuration control, policy enforcement, logging, monitoring, evaluation, and cost management. It also needs clear pathways for approving new use cases and retiring weak ones. A mature enterprise AI strategy is selective. It does not turn every experiment into infrastructure; it identifies the use cases with durable value and then builds the supporting capability properly.

Platform thinking also includes scaling capability through AI agent collaboration. You can build a small army of AI agents, each performing a useful task, but their greater value is unlocked when they collaborate to handle more complex, higher-value work while lowering unit costs.

Pilots matter, but only if they lead to learning or repeatable change. In the end, enterprise value comes not from the number of pilots launched, but from the number of capabilities robustly embedded in real business operations at scale.

Measurement, Reliability, and Board-Level Value

One danger of AI enthusiasm is that success becomes associated with novelty rather than evidence. A system may seem impressive because it produces fluent language, persuasive summaries, or elegant answers. Yet in enterprise settings, the important question is not whether a system looks intelligent, but whether it performs reliably, changes the right workflow, reduces the right cost, improves the right decision, or supports the right strategic objective. Without disciplined measurement, organizations can mistake activity for progress and experimentation for value.

Measurement in enterprise AI must operate at more than one level. There is the performance of the model or system itself, accuracy, relevance, completeness, consistency and compliance. There is also task performance, whether users complete work more quickly, quality improves, customer outcomes improve, throughput rises, and error rates decline. Finally, there is the economic and strategic dimension, whether the capability justifies continued investment, supports a priority business objective, and can be scaled without disproportionate cost or risk. This matters because a technically capable system can still fail as an enterprise investment if it does not improve outcomes and deliver business value.

Reliability deserves particular emphasis because it is so closely tied to trust. Enterprise users do not need perfection from AI, but they do need predictability that fits the context of use. A system that is very helpful nine times out of ten may still be unsuitable for a process in which the tenth failure is costly or dangerous. This means enterprises must design for imperfect reliability, clear use cases, confidence-aware workflows, escalation paths, human oversight, and monitoring of failure patterns in operation.

For boards and executive teams, however, trust and reliability must ultimately connect to value. That is where many AI initiatives struggle. They remain framed as innovation programmes or technology experiments rather than business capabilities with a clear investment case. The conversation needs to be translated into terms boards can act on, what value is expected, how it will be measured, what controls protect the business, what the implementation and operating costs are, and what risks remain. In this respect, the idea of AI as a digital employee can again be useful. It encourages leaders to consider productivity contribution, supervision cost, role fit, risk exposure, and return on investment in practical terms.

Measurement, reliability, and productivity are the disciplines that turn AI from an interesting possibility into a credible business proposition.

Building Secure, Scalable, and Trusted AI Systems

The most useful way to think about enterprise AI is not as a race to adopt the latest model, nor as a narrow technology deployment hidden within IT. It is better understood as an effort to build adaptive, trusted systems in which AI plays a controlled but powerful role. The copilot metaphor keeps this ambition grounded. A copilot assists rather than replaces and operates within a system of shared responsibility, procedure, and judgement. That is precisely what enterprises need if they are to benefit from AI without undermining security, trust, or operational control.

No framework guarantees success, but organizations are far better positioned to thrive during periods of high disruption when they select problems to solve meaningfully, build on sound data and architecture, design for security and governance, and support people through the change.

About the Author

Born in the UK and now settled in Ireland, Peter Rose is TEKenable's co-founder and Group CIO, with forty years' experience delivering enterprise software, technology strategy, and business transformation. An expert in AI, data, secure architecture, enterprise integration, and low-code platforms, he has held senior technology leadership roles at Deloitte, Vision Consulting, and others before co-founding TEKenable in 2002.

Peter has led mission-critical programmes across banking, insurance, energy, public sector, and other regulated industries, aligning technology investment with business outcomes, security, scalability and compliance. He is a trusted advisor to executive teams and boards, bringing strong governance, cybersecurity and risk management expertise, supported by professional accreditations, including Fellow of the Irish Computer Society, Chartered Engineer, M.Inst.D, CISSP, CISM, CSSLP, and CCSP.

A measured voice on AI and digital transformation, Peter champions responsible AI adoption as a copilot for people, helping organizations turn emerging technologies into sustainable business value.

Email: peter.rose@tekenable.com

Website: www.tekenable.com

LinkedIn Profile: https://www.linkedin.com/in/peterrose1/

CHAPTER 26

THE RECOMMENDATION AREA: FROM RANKING TO ENDORSEMENT

By Ghina Sahagun
Founder of FounderGate LLC
Boston, Massachusetts

We are the creative force of our lives, and through our own decisions rather than our conditions, if we carefully learn to do certain things, we can accomplish those goals.

—Stephen Covey

Throughout my career in information technology, I have enjoyed a front-row seat to several digital revolutions. Looking back, I can identify three distinct moments that fundamentally changed how businesses connect with customers, build trust, and drive growth.

The first was the rise of the website, which defined a company's online presence. The second was the explosion of social media, which reshaped how businesses built relationships and trust. The third is happening right now through artificial intelligence, which is redefining how businesses are discovered and recommended.

Each of these shifts initially appeared to be a pure technology story. In reality, they were trust stories. Technology simply altered how people discovered information, evaluated options, and made purchasing decisions. The businesses that succeeded were the ones that understood how customer behavior was evolving and adapted accordingly.

When websites first emerged in the 1990s, many business owners viewed them as an optional luxury. A solid local reputation, a listing in the Yellow Pages, local newspaper advertising, and word-of-mouth referrals seemed entirely sufficient. Customers had always found businesses that way, so leaders felt little urgency to change. (Back then, if someone told you to get online, you first had to listen to your modem negotiate peace terms with the internet.)

Gradually, however, consumer behavior shifted. People began researching products and services online via Google or Bing before making a phone call or stepping into a storefront. A website transformed from a competitive advantage into a basic business requirement. Companies that embraced this internet shift early captured immense visibility and market share, while those that waited found themselves difficult to discover.

The same pattern repeated itself with social media. Many business owners initially dismissed Facebook as a platform for students and LinkedIn as a digital résumé archive. Few anticipated that social platforms would become critical infrastructures for brand building, customer engagement, and reputational management. Yet within a relatively short period, consumers began expecting businesses to have an active social media presence.

Today, we are witnessing another transformation that may prove even more significant than the previous two shifts combined. Artificial intelligence is changing not only how businesses operate internally but also how customers search for information, evaluate competing options, and ultimately decide whom to trust.

For nearly three decades, search engines functioned like digital librarians. (It was an era where clicking through those links felt a bit like a digital treasure hunt.) Users entered a specific keyword query and received a list of websites that might contain the answer. The customer carried the burden of clicking through those links, comparing options, evaluating credibility, and making a final decision.

AI is fundamentally dismantling that experience. Users increasingly turn to AI to synthesize information rather than merely locate it. Instead of sorting through ten blue links and conducting extensive manual research, users ask AI systems to provide direct recommendations, summarize complex options, and identify the single best solution for their specific problem.

For business leaders, this represents a profound paradigm shift. Businesses are no longer competing solely for top search engine rankings. They are competing to become the trusted source that AI systems reference, cite, and present to users.

The question is no longer who ranks first. The question is who gets recommended. And increasingly, those recommendations are based on demonstrated expertise, credibility, and trust rather than visibility alone.

Why AI Search Changes the Rules of Visibility

We are moving from a world where search engines helped users find information to one where AI systems increasingly help them interpret it. Traditional search engines were designed to locate relevant sources. AI systems are designed to evaluate information, synthesize it, and provide a direct answer.

Importantly, this shift is not happening because consumers have suddenly changed the way they search. In many cases, they are still asking the same questions they always have. A homeowner who once typed "best flooring contractor Boston" into Google may now enter that exact same query into an AI assistant.

The difference is not necessarily in the question itself. It is in what happens next. Instead of presenting a list of links and requiring users to conduct their own research, AI systems increasingly attempt to interpret intent, evaluate options, and provide a synthesized response. As consumers become more comfortable interacting with AI, many are also beginning to ask more detailed and conversational questions. A homeowner may ask, "Who specializes in restoring original pine floors in historic New England homes?"

The direction is clear. Search is evolving from a process of finding information toward a process of obtaining answers. This evolution changes the nature of digital competition. Businesses are no longer competing solely to rank for keywords. They are competing to become the source that AI systems choose to include in their responses.

This is where generative engine optimization (GEO) enters the conversation. While traditional search engine optimization (SEO) focuses on helping search engines discover, index, and rank web pages, GEO focuses on helping AI systems understand, verify, and trust a business's expertise when generating responses. SEO was designed for a world of links. GEO is designed for a world of answers.

The distinction is important. SEO competed for clicks. GEO competes for inclusion. SEO focused heavily on individual pages and keyword relevance. GEO places greater emphasis on expertise, credibility, authority, and consistency across an organization's broader digital footprint.

Yet GEO is not a replacement for SEO. It is its next evolution. Businesses do not need to tear down the house they spent years building. They need to build on it. The same elements that contributed to online success in traditional searches remain highly relevant today. Authoritative content, positive reviews, quality backlinks, consistent business information, technical excellence, and demonstrated expertise continue to serve as critical signals of credibility. Organizations that have invested in SEO, content creation, and reputation-building have already completed much of the foundational work required for the AI era.

The foundation has not changed. What has changed is how it is evaluated. Success is no longer defined by where a business appears in search results but whether that business becomes part of the answer itself.

The Opportunity for the Newcomers

If established businesses hold the advantage of history, how can a new business hope to survive in an AI-driven marketplace? This is perhaps the most exhilarating paradox of the AI revolution: While AI deeply respects the old guard, it simultaneously creates an unprecedented fast-track for the new.

Historically, breaking into an established market required immense capital and years of patient grinding. Large, entrenched corporations could hire premier agencies, maintain massive marketing departments, dominate television and radio advertising, and produce content at a scale that smaller upstarts could never match. In the old system, volume and budget won.

AI is beginning to break that monopoly. Today, a new business is no longer at the same disadvantage simply because it lacks a large budget or an extensive marketing team.

AI does not automatically favor the biggest brand in the room. It favors the business that appears most qualified to solve the problem being asked about. As consumers increasingly describe their specific needs rather than relying on a few keywords, AI systems are looking for expertise rather than size. This trend is likely to accelerate as voice-based interactions become more common. Few people speak the way they type into a search engine. Instead of typing "music lessons Boston," a parent is more likely to ask, "Which music school near me has the best program for a teenager preparing for a college music audition?" Conversational AI is bringing searches closer to the way people naturally communicate.

This has given rise to the micro-authority: a business that deliberately chooses to be exceptionally trusted within a highly specific, tightly defined niche. Because they are starting today, new businesses can build their entire presence around this conversational era. They are not weighed down by outdated marketing habits, legacy websites, or generic content footprints. They can immediately focus their entire public identity on solving hyper-specific problems, allowing them to accelerate the process of building visibility and demonstrating expertise. In some cases, this enables them to earn recognition for capabilities that larger competitors may overlook or struggle to communicate effectively.

Why Trust Matters More Than Ever

Perhaps the most significant consequence of generative AI is that it fundamentally changes the value of information.

For much of the digital era, creating content required time, expertise, and human resources. Today, anyone can use AI to generate data in a matter of minutes. As content generation becomes effortless and practically free, the volume of noise on the internet is growing exponentially.

Consequently, information is rapidly declining in value as a differentiator. This creates a powerful operational paradox: At the exact moment information becomes abundant, trust becomes incredibly scarce.

Consumers are surrounded by an endless sea of automated, synthesized text. They no longer suffer from a dearth of information; instead, they suffer from a lack of confidence. When looking to hire a professional or purchase a service, they want absolute assurance that the facts they receive are accurate and battle-tested.

AI models face a similar challenge. They must continuously evaluate which sources deserve to be trusted as the volume of online content continues to expand.

This environment presents both a warning for the old guard and an open door for the new. Established businesses cannot coast on their old search rankings by churning out uninspired, AI-generated filler. New businesses cannot expect to win by simply clicking "generate" on a standard marketing tool.

The future will not belong to the organizations that produce the largest volume of generic content. It will belong to the organizations that provide the strongest, most verifiable evidence of human authority. Case studies, testimonials, proprietary data, and personal insights are the AI age's true currency.

The Advantage of Being Established and the Advantage of Being New

One of the most fascinating aspects of the AI era is that it appears to be rewarding two very different types of businesses at the same time. Established businesses bring history. New businesses bring agility. AI rewards both.

On one side are companies that have spent years, sometimes decades, building their reputation. They have accumulated reviews, testimonials, media mentions, backlinks, educational content, and customer success stories across the internet. Many have also invested heavily in SEO over the years, occasionally wondering whether all those blog posts, directory listings, and website updates were really worth the effort. As it turns out, they were doing far more than improving rankings. They were building a digital track record of expertise and trust that AI systems can now analyze in seconds.

That does not mean established businesses can put their feet up and assume AI will do the rest. History helps, but it is not a lifetime membership card. A company that relies on content written ten years ago and never adapts to changing customer questions may discover that AI has a surprisingly short attention span. Trust built over time is valuable, but it still needs to be maintained.

New and emerging businesses may not have decades of reviews or a library of content, but they possess something equally powerful: agility. They are not burdened by legacy processes or the phrase, "We've always done it this way." They can build their online presence specifically for the AI era, speaking directly to the detailed, conversational questions customers are asking today.

This is where specialization becomes a superpower. A large company may try to serve everyone, but AI is often looking for the most precise answer to a very specific question. A boutique firm that focuses on a narrow niche can sometimes earn a recommendation ahead of a much larger competitor simply because its expertise is clearer, deeper, and more relevant to the problem being solved.

AI is creating a marketplace where both experience and focus matter. Established businesses benefit from years of accumulated trust. Emerging businesses benefit from the ability to move quickly and carve out highly specialized areas of expertise. One has history on its side; the other has flexibility.

The real surprise is that AI is not creating a winner-takes-all environment. It is doing something much more interesting. It is rewarding credibility wherever it finds it. That may be the most democratic aspect of AI search. It respects the reputation built by established businesses while giving newcomers a genuine opportunity to compete. In the end, AI is asking, "Who is the best fit?"

Navigating the AI Transition

The path forward is actually simpler than it may appear. The goal is no longer just to be visible online. The goal is to become recommendable. Thriving in the AI era does not require abandoning proven business fundamentals. The same principles that built successful businesses before AI remain just as important today: expertise, trust, customer satisfaction, and willingness to adapt.

The first step is understanding how businesses currently appear in the AI landscape. Try asking: "What is this company known for?" Ask who the leading providers are in your industry and geographic area. Ask which businesses specialize in the specific problems you solve. The answers can be surprisingly revealing. Business leaders may discover that AI understands their business expertise perfectly. They may also uncover gaps between theirs and AI perception. Think of it as the digital equivalent of overhearing a conversation about yourself at the next table in a restaurant. Sometimes it is reassuring. Sometimes it is enlightening. Occasionally, it is a little uncomfortable.

Many organizations discover that the services they most want to be known for are not necessarily the services for which they are most visible online. In other cases, they may find that AI understands what they do, but not the full depth of their expertise. If a business's website, content, reviews, and online presence emphasize one aspect of expertise while underrepresenting another, AI will naturally reflect that imbalance.

These discoveries provide a roadmap for improvement. Businesses gain direct insight into where messaging, content, and authority-building efforts should be focused. An AI query can reveal opportunities that might otherwise have taken months of market research to uncover.

Business leaders should also think about the digital evidence their organization leaves behind. AI systems are constantly evaluating signals of credibility. Every positive interaction that leaves a digital footprint becomes another piece of evidence supporting expertise.

Leaders should resist the temptation to view AI as purely a marketing initiative. The organizations seeing the greatest benefits are using AI throughout their business. They are streamlining research, improving customer service, accelerating content creation, supporting decision-making, and automating repetitive tasks. The objective is not to replace people. It is to give people more time to focus on the work that creates the most value.

AI may change how customers discover businesses, but it does not change what customers ultimately value. They still want competence. They still want reliability. They still want results. The companies that succeed will be the ones that use AI to enhance those qualities rather than replace them. They will be the ones that successfully combine innovation with credibility, efficiency with authenticity, and artificial intelligence with human intelligence. Technology may help businesses get discovered, but trust remains the aspect that turns a recommendation into a relationship.

Human Connection in an AI-Driven World

What traits become significantly more valuable as automation expands? The answer is remarkably human.

Artificial intelligence can process, analyze, and synthesize data at staggering speeds. It can draft thousands of words of text in seconds, identify hidden patterns in consumer behavior, and streamline complex operational workflows. Yet it cannot build authentic human relationships. It cannot demonstrate genuine, lived empathy. It cannot earn trust through years of consistent, ethical real-world behavior, and it cannot navigate the deep emotional complexities that drive high-stakes human decision-making.

The businesses that thrive over the next decade will not make a binary choice between human talent and artificial intelligence. Instead, they will use AI as an efficiency engine to handle baseline execution, freeing up their people to invest more deeply in high-value activities: client relationship-building, creative problem-solving, strategic leadership, and empathetic communication. Technology will handle the execution of data; human beings will continue to provide meaning, judgment, and connection.

Conclusion: Visibility Evolves, Trust Endures

Unlike previous digital revolutions, AI is doing something unusual. It is rewarding both incumbents and challengers at the same time. Established organizations benefit from years of accumulated trust and authority, while emerging businesses benefit from dramatically lower barriers to building visibility and demonstrating expertise. In many ways, AI is creating the most meritocratic discovery system we have seen so far. It respects history, but it also rewards relevance.

History has repeatedly proven that while the tools of discovery evolve, the underlying psychology of the human buyer remains remarkably constant. The transition from the physical Yellow Pages to the search bar did not alter the fundamental truth that customers prefer to do business with organizations they trust. It merely changed how they found them.

The current transition to artificial intelligence follows this same pattern. Artificial intelligence is not an existential threat to authentic, high-performing businesses. It is a powerful filter. It separates superficial digital noise from expertise. It disadvantages organizations that rely on shortcuts, vanity metrics, and generic content, while amplifying businesses that have invested in knowledge, credibility, and genuine customer value. In the end, AI is making trust more visible, more measurable, and more valuable than ever before.

The businesses that thrive will be the ones that consistently demonstrate expertise, earn credibility, and deliver meaningful value to their customers. AI is simply becoming the newest way that trust is recognized, validated, and rewarded. The question is no longer who ranks first. The question is who gets recommended.

About the Author

Ghina Sahagun is the founder of FounderGate LLC, a consultancy specializing in digital marketing, AI visibility, and business transformation. She helps organizations strengthen their online presence through SEO, GEO, and strategic digital initiatives. With more than twenty-five years of experience spanning technology, market intelligence, and business strategy, she helps businesses navigate change and build sustainable competitive advantage.

Before founding FounderGate, Ghina spent more than a decade at IBM in France and the Netherlands. She holds an MBA from HEC Paris and a dual degree in mathematics and software engineering from Paris Dauphine University.

Email: gsahagun@foundergate.com

Websites: https://foundergate.com

https://ghinasahagun.com

LinkedIn: https://www.linkedin.com/in/ghinajabadosahagun/

CHAPTER 27

AI-OPTIMIZED INFRASTRUCTURE: DRIVING EFFICIENCY, RESILIENCE, AND INNOVATION AT SCALE

By Mohammed Asif Sheroz Ali
Principal Architect and AI Evangelist
Riyadh, Saudi Arabia

We are moving to a world where GDP growth of a country will be based on the amount of AI compute the country has.

—Bret Taylor

Every meaningful AI breakthrough you have ever heard about—ChatGPT answering a complex question, a medical AI detecting cancer earlier than a human doctor, a self-driving car navigating a busy intersection—was made possible by something invisible: an extraordinary foundation of compute power, high-speed networking, energy systems, and data architecture. Infrastructure is the silent engine of the AI revolution, and understanding it is not just for engineers. It is a strategic mindset every leader, entrepreneur, and professional must develop to stay relevant in the years ahead.

The Trillion-Dollar Moment: AI's Fastest Ascent in Economic History

Something unprecedented is happening in capital markets. AI-native companies are reaching trillion-dollar valuations at a speed that has no historical precedent.

Legacy technology giants—Apple, Microsoft, Amazon, Google—each required roughly twenty years of growth to cross the $1 trillion market capitalization threshold. The most advanced AI companies are achieving equivalent or greater valuations within three to five years of meaningful commercialization. This is not incremental acceleration. It is a categorical shift in how markets price transformative capability.

Several AI startups and pure-play AI platforms are now approaching or surpassing the trillion-dollar IPO valuation horizon, a benchmark that, even five years ago, seemed reserved for mature platform businesses with decades of customer relationships. The combination of generative AI adoption curves, enterprise SaaS conversion, and infrastructure monetization is compressing traditional value-creation timelines to a fraction of their historical norms.

The Road to $10 Trillion: The Next Generation of Market Giants

The next frontier is bolder still. As AI transitions from a productivity enhancement to the foundational operating system of the global economy, analysts and technologists are converging on a striking forecast: the top five companies by market capitalization within the next decade will overwhelmingly be AI-native enterprises, and the largest among them may approach $10 trillion in market value.

This projection rests on a simple but powerful thesis: AI is not a product category. It is an infrastructure layer, as foundational to 21st-century commerce as electricity was to the 20th. The companies that own the AI infrastructure stack—from silicon to models to agent platforms—will command value accrual at a scale the current generation of technology giants has never seen.

AI as Necessity: The New Mandatory Spend

We are approaching an inflection point: AI is transitioning from discretionary investment to mandated operational expenditure. Much as electricity and internet connectivity became baseline requirements for economic participation, AI subscriptions, model access, and agent platforms are becoming non-negotiable line items for organizations and individuals alike.

Within the next five to seven years, a meaningful portion of every working professional's income will be allocated to AI tools as a condition of competitive employability. Across five billion knowledge workers globally, this represents a multi-trillion-dollar recurring revenue base structurally insulated from traditional demand cycles.

The Scale That Defies Imagination

To appreciate where AI is heading, consider the numbers. In 2022, global AI compute capacity stood at roughly 1 ZettaFLOP, that is 10^{21} floating-point operations per second. By 2026, that figure has grown to approximately 100 ZettaFLOPS. By 2030, projections point to nearly 10 YottaFLOPS, yet another thousand-fold leap. In less than a decade, the world will transition from performing trillions to quadrillions of AI operations every single second.

For leaders and professionals, the critical insight is this: AI is not coming. It is already here, operating at civilizational scale. The question is not whether you will be affected. It is whether you are building the mindset and capacity to harness it.

Why Infrastructure Is a Leadership Issue, Not Just a Technical One

There is a common trap non-technical leaders fall into: treating infrastructure as someone else's problem. "That's IT's job," the thinking goes. This mindset is becoming dangerously outdated.

Consider the analogy of electricity in the early 20th century. When electrification transformed factories, the companies that thrived were led by visionaries who understood that electricity was not just a utility, but a strategic platform. It changed how work was organized, what products could be made, and how fast businesses could grow. Infrastructure shaped competitive advantage.

AI infrastructure is today's electricity. Organizations building scalable, AI-optimized compute environments are unlocking capabilities their competitors simply cannot access. They can train new models, run real-time inference at global scale, and experiment at a speed that compounds into decisive market advantages.

The Architecture of the AI-Powered Enterprise

So what does modern AI infrastructure actually look like? At its core, it is built around three interlocking principles: compute density, high-speed connectivity, and energy efficiency.

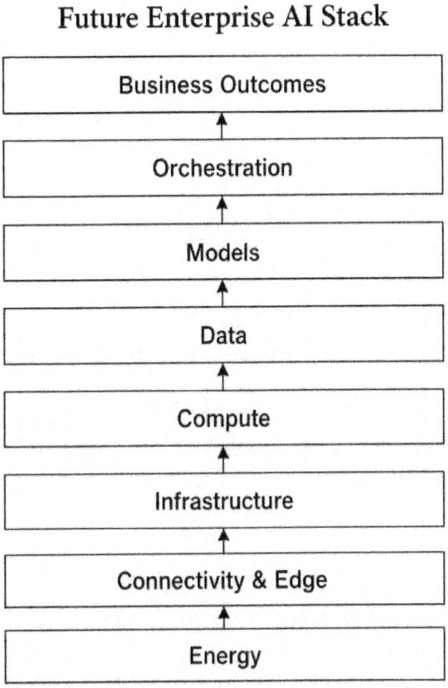

Figure 4. Future Enterprise AI Stack. Foundational layers enable higher-level intelligence, orchestration, and business decision outcomes.

Compute Density: The Power of Parallel Processing

Traditional computers process tasks sequentially. AI workloads are fundamentally different. They require massive parallel computation, executing thousands of operations simultaneously. This is why modern AI systems are built around GPUs (graphics processing units), NPUs (neural processing units), and specialized AI ASICs rather than conventional processors. Semiconductor technology is advancing toward the 1nm process node, enabling unprecedented transistor density and performance-per-watt gains that will define the next generation of AI silicon. The installed base of GPUs and AI accelerators is expected to expand from around a few million units in 2025 to approximately tens of millions by 2030, reflecting the explosive scaling of AI compute demand.

AI-specific silicon is now dominant. AI-specific silicon is projected to handle more than 90% of AI workloads by 2027, displacing general-purpose CPUs for all inference and training tasks. This silicon specialization is a direct driver of the performance-per-watt gains making large-scale AI economically viable.

Today's AI training clusters contain tens of thousands of these processors, interconnected through ultra-high-speed networking fabrics, collaborating on tasks that would be impossible for any single machine. The result is the ability to train foundation models containing hundreds of billions or even trillions of parameters.

Strategic Insight: Compute density is not just about raw power. It is about building environments where experimentation is fast and affordable, enabling organizations to rapidly test, train, and iterate on AI models.

High-Speed Connectivity: When Every Millisecond Counts

AI clusters do not think in isolation. They are distributed systems where thousands of processors must constantly communicate with near-zero delay. The next evolution in this space is already underway: Photonic interconnects—using light rather than electrical signals to carry data between nodes—are beginning to replace much of the electrical networking fabric in hyperscale AI infrastructure.

Photonic interconnects offer transformative advantages: bandwidth that scales with wavelength division multiplexing, latency approaching the physical speed of light, and dramatically lower energy consumption per bit transmitted. As AI clusters grow toward exascale configurations, electrical interconnects become a bottleneck that photonics is uniquely positioned to resolve.

For business leaders, the practical implication is a shift to latency-sensitive operations. As AI becomes embedded in customer interactions, financial transactions, healthcare diagnostics, and supply chains, the infrastructure must respond in milliseconds.

Energy Efficiency: The "Watts-per-Token" Imperative

There is no free lunch in AI infrastructure. Large AI clusters can consume tens to hundreds of megawatts of power, comparable to small cities. As global AI adoption grows from billions to potentially five billion users by 2032, energy consumption becomes not just an environmental concern, but a primary business risk.

"Watts per token" is emerging as a defining KPI for AI infrastructure efficiency, measuring the energy cost of generating each unit of AI output. Forward-looking organizations are already benchmarking this metric across model architectures, hardware generations, and deployment configurations.

The next generation of AI infrastructure is addressing this through liquid cooling, AI-driven energy optimization, renewable energy integration, and advanced power distribution architectures. AI infrastructure rack power densities are expected to increase substantially, from roughly 50 kW per rack today to 250 kW liquid-cooled racks, driven by platforms pushing the boundaries of compute per footprint. Global data center capacity is projected to grow from approximately 80 GW in 2025 to nearly 290 GW by 2030, underscoring the scale of energy infrastructure investment required. Organizations and nations that invest in sustainable AI infrastructure will gain a dual advantage: lower operating costs and stronger alignment with ESG expectations.

From Centralized Hubs to Distributed Intelligence Ecosystems

Perhaps the most architecturally significant transformation in AI infrastructure is the shift from centralized compute—the hyperscale data center model that has defined the cloud era—to distributed intelligence ecosystems embedded everywhere.

AI infrastructure will evolve from centralized compute hubs into distributed intelligence ecosystems embedded everywhere: in networks, edge devices, vehicles, industrial equipment, and personal technology. This is a fundamental reimagining of where intelligence lives and how it is delivered.

The Defining Characteristics of Distributed AI Infrastructure

The distributed intelligence era is defined by a set of architectural principles that represent a clean break from the centralized paradigm:

- **Near-Zero Human Intervention in Scaling Decisions:** Infrastructure scales autonomously in response to demand signals, without requiring human operators to provision, configure, or approve capacity changes.
- **Automatic Workload Decomposition Across Heterogeneous Compute:** Workloads are automatically analyzed and decomposed, routing inference tasks to edge silicon, training to GPU clusters, and lightweight operations to CPU-based endpoints.
- **AI-Specific Silicon Dominates >90% of Workloads:** GPUs, NPUs, and AI ASICs become the default substrate for virtually all AI workloads, with general-purpose compute reserved only for orchestration and legacy compatibility.
- **Photonic Interconnects Replace Electrical Networking:** Light-based data transmission replaces much of the electrical networking fabric, enabling the bandwidth and latency profiles required for distributed inference at planetary scale.
- **Millions of Persistent AI Agents per Enterprise:** Large enterprises deploy millions of purpose-built AI agents operating continuously across functions—finance, operations, customer experience, supply chain, R&D—each maintaining a persistent state.
- **"Watts per Token" as a Primary KPI:** Energy efficiency at the inference layer becomes a board-level metric, driving

hardware procurement, model architecture selection, and data center investment decisions.

- **Edge Devices Run Local Intelligence:** Phones, industrial sensors, medical devices, and vehicles run substantial AI models locally, enabling real-time decision-making without round-trip latency to central infrastructure.
- **Unified Control Plane for Compute, Data, Agents, and Security:** A single orchestration layer manages the full stack—compute allocation, data pipeline management, agent lifecycle, and security policy enforcement—across cloud, on-premises, and edge environments simultaneously.

For enterprise leaders, this shift has immediate strategic implications. The question is no longer "How much cloud capacity do we need?" but rather "How do we design a distributed intelligence fabric that places AI capability precisely where needed, at the lowest possible energy and latency cost?"

AI Running on AI: The Self-Optimizing Frontier

AI is now being used to manage the very infrastructure it runs on. Modern cloud platforms deploy AI-driven systems to monitor, predict, and optimize their own operations in ways human administrators cannot match.

Self-Optimizing Capabilities

- **Predictive Capacity Management:** AI models analyze historical workload patterns to forecast future demand, enabling dynamic resource allocation before bottlenecks occur.
- **Intelligent Workload Scheduling:** AI-driven schedulers determine optimal placement of compute tasks across clusters, maximizing performance and utilization.

- **Automated Fault Detection:** Machine learning systems continuously analyze telemetry data and hardware signals to identify anomalies and predict failures before they cause downtime.
- **Energy Optimization:** AI algorithms dynamically manage cooling, power routing, and workload placement to minimize energy consumption without sacrificing performance.

Infrastructure as an Innovation Platform: Industry Transformations

AI infrastructure is not a cost center. It is an innovation platform. Organizations that build or access scalable AI infrastructure unlock entirely new capabilities, not just incremental improvements, but fundamentally new possibilities across every industry.

Healthcare

AI infrastructure supports real-time medical imaging analysis, accelerated drug discovery simulations, and personalized treatment recommendations at a scale and speed no human team could achieve. Infrastructure is not just supporting healthcare. It is actively saving lives.

Finance

Financial institutions leverage AI infrastructure for real-time fraud detection, advanced risk modeling, and algorithmic decision-making. The institutions with the fastest, most capable infrastructure make better decisions faster, translating directly into competitive advantage and reduced risk.

Telecommunications

Telecom operators use AI to optimize network performance in real time, predict equipment failures before they happen, and manage the extraordinary complexity of modern digital ecosystems. Infrastructure has become the product.

Scientific Discovery

High-performance AI infrastructure is accelerating breakthroughs in climate modeling, materials science, genomics, and space exploration, solving problems that would have taken decades using traditional computing.

The Road to YottaFLOPS: What Comes Next

Between 2023 and 2027, AI compute performance is projected to grow by more than 1,000 times, a transformation in the nature of what is computationally possible.

Converging Innovations

- **Specialized AI Hardware:** New generations of AI accelerators designed specifically for deep learning workloads are delivering dramatic gains in performance per watt. Platforms such as NVIDIA's Kyber-based NVL576—featuring 576 GPUs per rack and anticipated around early 2028 with 600 kW power requirements—exemplify the trajectory toward extreme compute density within a single rack footprint.
- **Distributed Training Architectures:** Advanced techniques allow massive AI models to be trained across thousands of processors simultaneously, without loss of efficiency.

- **Software Optimization:** AI frameworks are improving their utilization of hardware, squeezing more capability from existing infrastructure.
- **Quantum and Neuromorphic Research:** Emerging computing paradigms may unlock entirely new dimensions of AI capability in the coming decade.

The threshold of 10 YottaFLOPS by 2030 is not just a benchmark. It represents a level of computational power capable of addressing some of humanity's most complex challenges, from climate change modeling to disease eradication.

Resilience: The Unsung Requirement of AI-Dependent Enterprises

As AI becomes woven into critical systems—healthcare diagnostics, financial markets, energy grids, transportation—the stakes of infrastructure failure rise dramatically. Resilience is no longer optional. It is a fundamental requirement.

Multi-Layered Resilience Architecture

- **Multi-region cloud architectures** that survive localized failures
- **Fault-tolerant distributed systems** designed to continue operating even when components fail
- **Cybersecurity frameworks** that protect AI models from adversarial attacks and data poisoning
- **Data integrity and governance structures** ensuring AI decisions are based on trustworthy information
- **Sovereign AI infrastructure** for organizations and nations requiring data privacy and national security assurance

For leaders, resilience thinking must extend beyond IT. When AI becomes central to operations, a service outage is not simply an inconvenience. It is an organizational crisis. Organizations that thrive in the AI era will invest in resilience as seriously as capability.

Your Competitive Position in the Infrastructure Era

You may be reading this as a technology executive, entrepreneur, coach, consultant, or a professional in a field that seems far removed from data centers and GPU clusters. Let me be direct about why this matters for you.

AI infrastructure is becoming a dividing line. On one side are individuals and organizations who understand what AI can and cannot do. They grasp why some implementations succeed and others fail. On the other side are those who view AI as magic: unpredictable, mysterious, and something that happens to them rather than something they can shape.

Three Mindset Shifts That Can Immediately Elevate Your Position

Shift 1: Think in platforms, not tools. Every AI capability you use is built on infrastructure. The leaders who win are not those who adopt the best tools. They are those who build or access the best platforms. Ask not "What AI tool should I use?" but "What AI platform gives me sustainable, scalable advantage?"

Shift 2: Invest in infrastructure fluency. You do not need to become a systems engineer. But you do need to speak the language well enough to ask the right questions, evaluate vendors credibly, and lead your organization through intelligent AI investment decisions.

Shift 3: Prioritize resilience alongside capability. In the rush to adopt AI, many organizations neglect the foundational question: "What happens when this fails?" Build AI systems worthy of trust by building them on infrastructure worthy of trust.

Building the Foundations That Will Define the Future

The AI revolution is not simply about smarter algorithms. It is about building infrastructure capable of supporting intelligence at planetary scale.

From 1 ZettaFLOP in 2022 to a projected 10 YottaFLOPS by 2030, global computing power is expanding at a pace that has no historical precedent. AI users are growing from millions to billions. AI startups are reaching trillion-dollar valuations in a fraction of the time it took legacy technology giants. The top five companies by market cap within a decade will likely be AI-native enterprises, with the largest approaching $10 trillion in value. And every working professional, across every sector and geography, will soon allocate a portion of their income to AI access as a condition of economic participation.

The infrastructure supporting all of this is simultaneously evolving from centralized compute hubs to distributed intelligence ecosystems embedded everywhere and characterized by near-zero human intervention in scaling, automatic workload decomposition, AI-specific silicon dominance, photonic interconnects, millions of persistent agents per enterprise, watts-per-token efficiency metrics, edge-native intelligence, and unified control planes spanning compute, data, agents, and security.

The organizations and nations that invest strategically in AI-optimized, energy-efficient, resilient, distributed infrastructure will define the next era of economic productivity, scientific leadership, and human progress.

The future of AI will not be built solely in algorithms or lines of code. It will be built in data centers, networks, silicon, photonic fibers, and the decisions made by people like you: leaders willing to understand the foundations of civilization's next great leap, and bold enough to build on them.

Physical AI will extend AI into the real world, increasing the need for scalable, high-performance infrastructure capable of supporting autonomous decision-making, simulation, perception, and real-time control.

The question is not whether AI will transform your world. It already is. The question is whether you will be among those who shaped that transformation or simply those who experienced it.

About the Author

Mohammed Sheroz is a technology leader, Principal Architect (EA) with over fifteen years of experience in IT infrastructure, AI Infrastructure, cloud computing, enterprise architecture, and digital transformation. A graduate of Osmania University in electronics and communication engineering, he began his career at HPE, leading large-scale infrastructure transformation programs for global enterprises such as PepsiCo and BOA. He has held expertise in DC design, migration, implementation, and operations of IT infrastructure, as well as the enablement of public cloud platforms, including Microsoft Azure and Oracle Cloud in KSA.

Currently serving as principal architect at Saudi Telecom Company (stc), Sheroz has been instrumental in driving cloud and AI transformation, EA and DC strategy. His achievements include establishing stc's first strategic cloud partnership with Oracle. Leading cloud governance initiatives that delivered over $300 million USD in infrastructure cost optimization, and advancing AI-ready DC and private cloud strategies.

He has also been instrumental in accelerating cloud adoption, transforming the deployment of open-source DB by 200% and developing EA blueprints, technology strategies, and transformation roadmaps that enable sustainable innovation and future-ready digital platforms. His leadership and innovative initiatives contributed to stc's recognition as a runner-up in the Forrester EMEA EA Awards 2025.

Email: sherozscholar@hotmail.com

LinkedIn: https://www.linkedin.com/in/sherozscholar/

CHAPTER 28

THE ARCHITECTURE OF INSIGHT: BUILDING SCALABLE DATA FOUNDATIONS FOR AI

By Raymond Sondak
Founder, Data & Solution Architect, Co-author, Speaker
The Hague, Netherlands

> *Data-centric AI is the discipline of systematically engineering the data needed to successfully build an AI system...The dominant paradigm over the last decade was to download the data set while you focus on improving the code...It's now more productive to hold the neural network architecture fixed, and instead find ways to improve the data.*
>
> —Andrew Ng[1]

In 2012, MD Anderson Cancer Center partnered with IBM on what felt like the future arriving early. The ambition was to deploy Watson AI to help oncologists make faster, evidence-based treatment decisions, with MD Anderson providing the clinical expertise and IBM the technology.

Four years and approximately $62 million later, the project was terminated. No clinical tool was deployed. No patients were helped.[2] The case has been studied from many angles, but the data architecture lens has received the least attention. That is where this chapter focuses.

An audit by the University of Texas System Audit Office documented what went wrong.[3] The audit documented no mechanism for keeping training data current as clinical guidelines evolved. Nothing flagged it had fallen behind. Watson wasn't malfunctioning. It was doing exactly what it had been built to do. IBM and MD Anderson did not fail because the AI wasn't powerful enough. They failed because they built sophisticated intelligence on an ungoverned, unversioned, semantically rigid foundation, and were surprised when it did not hold. The data architecture was the failure.

The audit points to no versioning, no lineage, no governance to detect when the underlying data had become unreliable.

When AI projects fail, the autopsy almost always leads back to the data architecture, not the model. RAND's 2024 research found more than 80% of AI projects never reach production deployment, twice the failure rate of traditional IT projects.[4] Gartner predicted 60% of AI projects without AI-ready data will be abandoned, with 63% of organizations lacking adequate data management practices for AI.[5] MIT's NANDA initiative found 95% of organizations reported zero measurable financial return from generative AI.[6]

The failure is almost never the model. It's almost always the foundation beneath it. After all, AI is not magic. It is architecture. The organizations that understand this, and invest accordingly, are the ones generating real, sustained, measurable value from AI.

The Insight Gap

Organizations have more data than ever. Strategic clarity hasn't kept pace. Petabytes of records accumulate while leaders still make critical decisions on intuition, incomplete reports, and competing spreadsheets that disagree on basic facts. This isn't a data scarcity problem. It's a data architecture problem.

The insight gap, the distance between data volume and decision intelligence, has three structural causes that organizations consistently misdiagnose.

Cause 1: Semantic Fragmentation: "Revenue" means recognized revenue in the finance warehouse, booked contracts in the CRM, and gross transaction value in the product analytics platform. Without a shared semantic layer, any analysis crossing system boundaries becomes unreliable. Analysts spend most of their time reconciling definitions that should have been set at the platform level. AI systems trained on semantically fragmented data produce confident-sounding nonsense.

Cause 2: Lineage Opacity: Most organizations can't tell you where their data came from, and without that trail, trust collapses. A business user burned once by a wrong report stops trusting reports, then models. Nobody can explain where the numbers came from, and that's not a technology problem.

Cause 3: The Dashboard Illusion: A well-designed dashboard showing last quarter's performance isn't an analytical capability. It's a rearview mirror. Real insight means asking questions that haven't been asked, modeling outcomes that haven't occurred, detecting patterns no human would think to look for. None of that is possible without a foundation that makes data machine-interpretable, not just human-readable.

The takeaway: Closing that gap requires better design, not better tooling.

Architecture Before Algorithms

When an AI initiative fails, and most do, the autopsy focuses on the model. Wrong algorithm. Too little training data. Vendor overpromised. These diagnoses are almost always incomplete. The root cause usually isn't the algorithm. It's the infrastructure feeding it. RAND's 2024 research identifies data-driven failures as the second most common root cause of AI project failure.

What does a machine learning model need to function reliably in production? Historically consistent, complete, and unbiased training data. Feature inputs calculated identically during training and inference. Governance controls on permissible data use, and lineage so every output can be traced. Not one of these requirements is algorithmic. Every single one is architectural.

Dimension	Algorithm-First	Architecture-First
Primary investment	Model selection and tuning	Data platform and pipelines
Definition of quality	Clean enough to run	Governed, traceable, and trusted
Failure mode	Confident wrong answers	Delayed launches, durable outcomes
Scalability	Project-by-project	Platform-wide, reusable
Governance	Compliance exercise	Engineered into the platform
AI output	Black-box output	Decisions you can trace and defend
Time to value	Fast prototype, slow production	Slower prototype, durable production

The algorithm-first organization gets to a demonstration quickly. Then it hits production, where the data is messy, inconsistent, and poorly governed, and the project stalls. The gap between prototype and production isn't a machine learning problem. It's a data engineering problem.

The architecture-first organization builds lineage infrastructure before model development, shared semantic standards before anyone runs a query. When it deploys a model, it deploys into a platform built to sustain it, not pipelines held together by a team working harder than they should.

Ontology and Semantic Modeling

Ask a data engineer what makes AI unreliable and you'll hear about data quality: missing values, duplicate records, outliers. Real problems, but symptoms of something deeper most organizations never address. There's no shared, machine-readable understanding of what the data means. That's the problem ontology and semantic modeling solve.

Core Vocabulary

Ontology: A shared, machine-readable vocabulary of your business. It defines what concepts exist, how they connect, and what they mean, so humans and AI systems interpret data consistently.

Semantic Model: A consistent vocabulary that translates the physical structure of data (tables, columns, keys) into business-meaningful concepts with agreed definitions and explicit relationships.

Knowledge Graph: An organized network of real-world entities and the relationships between them. Both humans and AI systems can navigate it to understand how concepts connect across domains.

Without an ontology that represents semantic equivalence across systems, an AI model works with fragments of reality: different systems, different terms, no way to reconcile them. A governed semantic foundation lets AI reason across systems and produce outputs that reflect a coherent view of the domain.

When an ontology is populated with real organizational data, it becomes a knowledge graph: a connected structure AI agents can traverse to reason across domains and resolve ambiguous terms.

The Metadata-Driven Federated Lakehouse

MD Anderson spent $62 million and four years on an AI system that couldn't do five things: version training data, trace outputs to source, enforce semantic consistency, prevent degraded data from reaching the model, or enforce governance in practice.

That list is also a blueprint for the right foundation, and the answer isn't a product. It's a pattern, which I call the metadata-driven federated lakehouse. The metadata-driven federated lakehouse isn't a standard lakehouse or a data mesh. It's a synthesis that takes what works from both and adds what neither provides.

The standard data lakehouse delivers medallion layer structure, ACID transaction guarantees, time-travel versioning, and centralized governance.[7] Transparency is where it falls short. Transformation logic lives in code, readable by engineers but invisible to everyone else. In most lakehouses, documentation and transparency are not the same thing.

The data mesh distributes ownership to the domain teams who best understand their data, but leaves structure undefined.[8] No mandatory layer discipline, no enforced semantic foundation, no consistent quality mechanism across domains. Domain autonomy is the mesh's greatest strength, and also where implementations most commonly break down.

The metadata-driven federated lakehouse combines the structural discipline of the lakehouse with the domain ownership of the mesh, adding what makes both viable at AI scale: metadata as the operating system of the entire platform. Every ingestion pipeline, every transformation rule, every business definition, every quality threshold, every access policy, and every retention constraint is expressed as metadata, not as hardcoded scripts, not as policy documents. The platform has no black boxes. Not because documentation was written. Because the architecture makes black boxes structurally impossible.

The metadata-driven federated lakehouse isn't a published framework or an industry term. It's a name I've given to an architectural pattern from practice, refined through experience across different organizations and sectors. What I haven't found documented as a unified pattern is this: metadata as the execution engine across a precise five-layer medallion structure (following the medallion architecture introduced by Databricks, 2020), federated gold ownership as the default, and a maturity path from central governance to domain self-service without a platform rebuild.

Bronze raw is the immutable record of what arrived, when, and from where. Source data lands in its native format, unaltered: no renaming, no conversion, no business logic, no cleaning. A permanent, append-only archive of what source systems reported.

Bronze formatted converts that same data into Delta Lake or Apache Iceberg without applying business logic. Time-travel versioning begins here. Every version is retained and queryable, so you always know what version trained the model. All subsequent layers use the same format, ensuring consistent time-travel versioning, schema evolution, and queryability.

Silver cleansed and conformed is where raw source data becomes governed organizational data. Quality rules are applied: completeness checks, range validations, referential integrity. Data that fails is quarantined. Business definitions are enforced: agreed definitions of each business entity, standardized codes, resolved cross-system identifiers. This is the layer that would have prevented Watson's semantic fragmentation at MD Anderson, with one definition of "treatment episode" enforced before any of it reached the model.

Silver historized preserves the complete change history of every record, every state retained and queryable with a precise timestamp. This enables temporal AI: models that reason about change over time, not just current state.

Gold has two structural sub-layers. Gold dimensional organizes data into star or snowflake schemas for BI tools, self-service analytics, and reporting. Its interpretable structure supports explainable AI. Gold one big table provides denormalized, wide datasets optimized for high-performance queries and ML model training.

Gold: The Ownership Principle

Gold is federated by default. Domain ownership isn't optional or centrally assigned; it's the default accountability structure. Finance teams encode their expertise in the finance gold model, operations engineers in theirs. As long as silver data is available, domain teams build and publish gold models without waiting for a central queue. If silver isn't yet available, the platform delivers it rapidly, because onboarding a new source is a metadata configuration exercise, not a development project.

In a standard lakehouse, transformation logic is code. In this architecture, it's metadata. When it lives in code, only engineers can read it. When it's metadata, business analysts, regulators, and AI agents can all read it without a technical interpreter.

That's why IBM's engineers couldn't explain Watson's recommendations to the clinical team. The transformations required engineering knowledge to read. In this architecture, every transformation is a metadata record, and tracing where any dataset came from is a query, not an investigation.

The metadata model makes governance automatic. Access policies, retention constraints, and quality thresholds are enforced by the same execution engine that runs the transformation pipelines. No separate governance system, no compliance checklist. Governance happens every time data moves through the platform, whether anyone is watching or not.

Most organizations start at stage 1, with the central team owning everything. Stage 2 shifts gold ownership to domain teams as capability grows. By stage 3, domains define their own ingestion metadata, and the platform scales faster than the team.

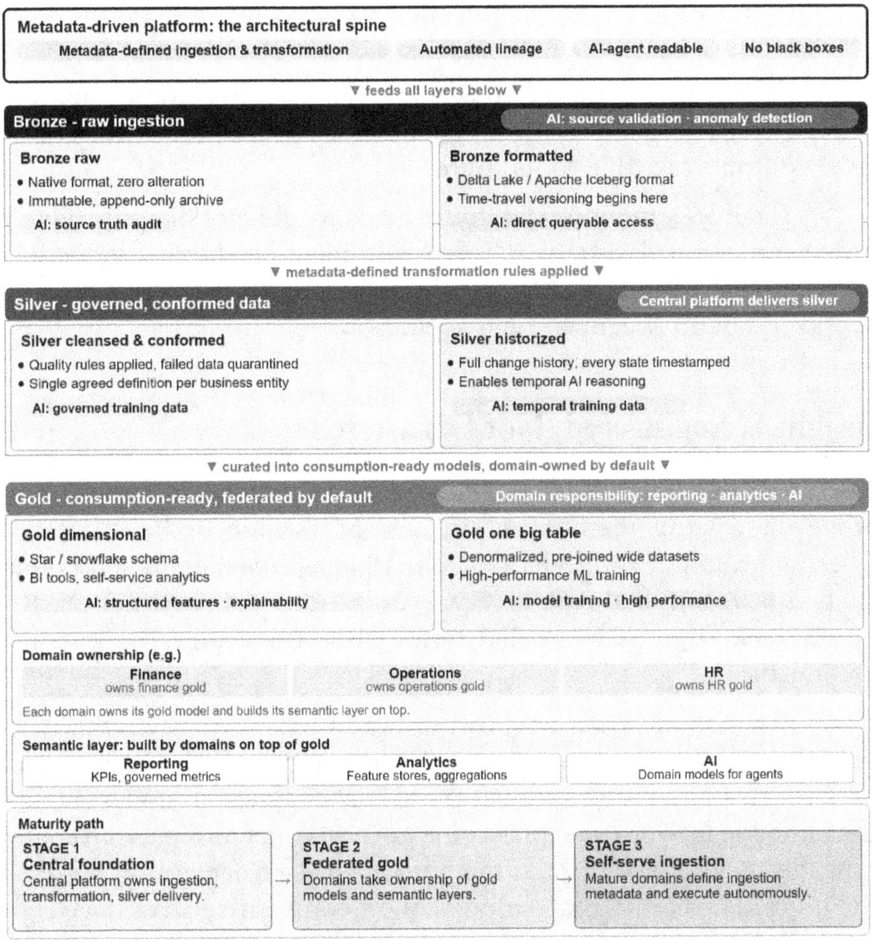

Figure 1: The Metadata-Driven Federated Lakehouse (Raymond Sondak, Analyticsaholic).

Built on a metadata-driven federated lakehouse, the MD Anderson story would have ended differently. Clinical guidelines would have landed in silver historized with a timestamp on every update. When the data product version changed, model retraining would have triggered automatically. Silent decay would have been structurally impossible. The semantic fragmentation across partner institutions wouldn't have happened either, with one definition of "treatment episode" enforced at silver.

When a regulator asked why the AI reached a recommendation, the answer would have been a metadata query, not an investigation. Every output traceable to its source.

The $62 million and four years would not have produced a decommissioned system. They would have produced a governed, versioned, fully traceable AI capability, where trust was an architectural property, not an organizational aspiration.

Insight Is Engineered, Not Discovered

There's a mythology embedded in how organizations talk about data: Somewhere within their data estate, insight is waiting to be discovered. The right analyst, the right tool, the right moment. This mythology isn't just wrong. It's expensive. Organizations that believe it invest in analysts. Organizations that know insight is engineered invest in architecture.

Two models define how organizations approach this. The heroic model relies on individuals who carry knowledge of which data to trust and how to transform it. The engineered model encodes that knowledge into the platform: governed definitions, automated pipelines, and data trails, so the answer doesn't depend on who's in the office. The heroic analyst looks like a competitive advantage right up until the day they leave.

The question isn't "Can this analyst figure it out?" It's "Has the architecture been designed to make this question answerable reliably and at scale?"

Dimension	Heroic Model	Engineered Model
Metric definitions	In analysts' heads	In a governed semantic layer

Dimension	Heroic Model	Engineered Model
Data quality enforcement	Manually, by experienced team members	Automated quality gates in the pipeline
New report lead time	Days to weeks	Hours, using reusable governed data products
Source traceability	Sometimes, with effort	Always, automatically, via lineage
Analyst departure risk	Institutional knowledge walks out the door	Encoded in the platform, not the person
Model degradation detection	When someone complains	Automated monitoring and alerting
New consumer onboarding	Needs veteran guidance	Data catalog provides context and confidence

The path from heroic to engineered is a discipline problem, not a technology problem. It means treating data as a product: owners, quality standards, versioning, and SLAs. It means defining insight precisely: not "we need analytics" but "we need to answer specific questions, with trusted data, traceable to its source."

The organizations that make this shift stop asking "What insight can we extract from our data?" and start asking "What architecture and foundation do we need to turn these questions from heroic effort into routine capability?"

Governance as Runtime, Not Afterthought

Walk into most organizations and ask where their data governance documentation lives. You'll be directed to a SharePoint folder, a Confluence page, or a PDF last updated two years ago. Ask who enforces it. You get a pause. Ask how. You get a longer pause.

Most organizations treat governance as a paper exercise: policies written by committees, reviewed periodically, and rarely enforced in practice. In an environment where AI makes decisions at speed and scale, that approach doesn't hold.

Dimension	Traditional Governance	Runtime Governance
Where policy lives	Documents and wikis	Metadata model and pipeline configuration
Enforcement	Manual review	Automated gates
Frequency	Quarterly or annually	Continuous, every data movement
Lineage tracking	Retrospective, when audited	Real-time, automated
Quality enforcement	Pre-deployment checklist	In-pipeline validation with quarantine
Access control	Granted by request, reviewed periodically	Policy-as-metadata, enforced at query time
Response to breach	Discovered after the fact	Blocked before it occurs
AI traceability	"We believe the data came from..."	Full audit trail to source, timestamp, and version

In the metadata-driven federated lakehouse, governance rules are metadata. Access policies, retention constraints, quality thresholds, and data trail requirements are expressed in the same metadata model as the pipelines. They're enforced automatically at every layer transition and data access event.

When a regulator asks whether the organization has complied, the answer is an audit log, not a reconstruction. When a privacy regulation changes, the metadata model is updated and the architecture enforces it automatically.

This is what separates AI experimentation from AI capability. Runtime governance doesn't slow organizations down. It removes the real bottlenecks: approval queues, manual reviews, compliance checklists that exist but never get enforced.

Practical Guidance

The architecture argument is only useful if it changes something. What follows is specific: one set of actions for individuals and one for organizations.

For Individuals

The professionals who will thrive are those who think architecturally about data.

Architectural thinking means asking questions most people aren't trained to ask: Where does this data come from? What assumptions are encoded in how it was collected? What governance constraints apply? What happens to this AI output if the upstream data changes? These aren't engineer questions. They protect every organization from confident decisions on corrupted foundations.

Action	Why It Matters
Map one AI system you interact with back to its data source. Ask where the training data came from, how old it is, and what governance applies.	The answers reveal more about trustworthiness than any benchmark.

Action	Why It Matters
Learn one concept from data architecture vocabulary: lineage, ontology, data product, or policy-as-code.	Building architectural vocabulary builds architectural thinking.
Identify the heroic analyst in your organization, the person whose departure would cause the most disruption, and ask whether their knowledge is encoded in the platform.	If it isn't, that's a risk worth surfacing.

For Organizations

These principles are not a checklist. They are the decisions that separate organizations that experiment with AI from those that scale it.

Principle	Concrete Action	Common Mistake to Avoid
Treat data architecture as a strategic enabler, not an IT expense	Include data architecture decisions in strategic planning, not only in sprint planning	Delegating architecture decisions entirely to technology teams without business leadership engagement
Adopt modular platforms grounded in shared ontologies	Before selecting any AI tooling, agree on a single definition for your most important business concepts	Selecting tools first and reconciling semantic assumptions afterward

Principle	Concrete Action	Common Mistake to Avoid
Build automated governance and quality pipelines before urgently needed	Identify the three most critical pipelines and implement quality gates, lineage, and access controls first	Attempting a comprehensive governance program that covers everything theoretically and enforces nothing practically
Treat AI as a continuous capability, not a project	For every AI initiative, define the operational model at the outset: ownership, degradation detection, retraining triggers	Treating model deployment as the finish line, then discovering production AI requires more ongoing investment than anticipated
Enable cross-functional teams where business, data, and technology work as one	Define the next AI initiative's team as cross-functional from day one, not as a data science project with business input	Treating cross-functional collaboration as a communication practice, not a structural decision

The Foundation Is the Future

The AI transformation is real. Its potential to reshape industries, accelerate discovery, and augment human judgment is visible in organizations that have done the work. The gains will go to those who build trusted, scalable, governed data foundations from the start, not those who move fastest.

The most valuable skill in the AI era isn't prompt engineering or model selection. It's thinking architecturally: asking where data comes from, what it means, how it's governed, and whether the system producing it can be trusted.

Organizations face a harder truth. Your AI strategy is only as credible as your data strategy. If you don't know where your data comes from, you can't trust your AI. If your governance lives in a document rather than the platform, you have a policy, not a control. If your semantic definitions live in individual heads rather than a shared model, your AI systems interpret the same reality differently, and that gap will show up.

The organizations that will lead the next decade are building right now. Not the best model, but the best foundation. After all, AI is not magic. It is architecture. And architecture, done right, is the most durable competitive advantage an organization can build.

References

1. Eliza Strickland, "Andrew Ng: Unbiggen AI," IEEE Spectrum, February 9, 2022, https://spectrum.ieee.org/andrew-ng-data-centric-ai.
2. Charlie Schmidt, "M. D. Anderson Breaks with IBM Watson, Raising Questions about Artificial Intelligence in Oncology," *JNCI: Journal of the National Cancer Institute* 109, no. 5 (May 2017): djx113, https://doi.org/10.1093/jnci/djx113.
3. University of Texas System Audit Office, *Audit of MD Anderson Cancer Center Oncology Expert Advisor Project* (Austin: University of Texas System, 2016), discussed in Eliza Strickland, "How IBM Watson Overpromised and Underdelivered on AI Health Care," *IEEE Spectrum*, April 2, 2019, https://spectrum.ieee.org/how-ibm-watson-overpromised-and-underdelivered-on-ai-health-care.

4. James Ryseff, Brandon F. De Bruhl, and Sydne J. Newberry, *The Root Causes of Failure for Artificial Intelligence Projects and How They Can Succeed: Avoiding the Anti-Patterns of AI* (Santa Monica, CA: RAND Corporation, 2024), https://www.rand.org/pubs/research_reports/RRA2680-1.html.
5. Gartner, "Lack of AI-Ready Data Puts AI Projects at Risk," press release, February 26, 2025, https://www.gartner.com/en/newsroom/press-releases/2025-02-26-lack-of-ai-ready-data-puts-ai-projects-at-risk.
6. Aditya Challapally, Chris Pease, Ramesh Raskar, and Pradyumna Chari, *The GenAI Divide: State of AI in Business 2025* (MIT NANDA, July 2025), https://mlq.ai/media/quarterly_decks/v0.1_State_of_AI_in_Business_2025_Report.pdf.
7. Michael Armbrust, Ali Ghodsi, Reynold Xin, and Matei Zaharia, "Lakehouse: A New Generation of Open Platforms That Unify Data Warehousing and Advanced Analytics," in *Proceedings of the 11th Conference on Innovative Data Systems Research* (CIDR, 2021), https://www.cidrdb.org/cidr2021/papers/cidr2021_paper17.pdf.
8. Zhamak Dehghani, *Data Mesh: Delivering Data-Driven Value at Scale* (Sebastopol, CA: O'Reilly Media, 2022).

About the Author

Raymond Sondak is a data architect, author, and consultant specializing in metadata-driven data architectures, AI-ready platforms, and digital transformation. He holds an MSc in media and knowledge engineering (AI focus) from Delft University of Technology and is co-author of *The Biml Book* (Apress). As founder of Analyticsaholic in the Netherlands, he advises data leaders and transformation-driven organizations on scalable, governed data ecosystems that enable responsible AI adoption. Raymond brings deep implementation experience in modern data platform design, semantic modeling, data governance, and organizational change. His work bridges technical innovation and business value, grounded in the conviction that AI amplifies operational maturity, for better or worse. He speaks at industry conferences and coaches leaders on AI strategy and data platform modernization.

Email: raymond.sondak@analyticsaholic.com

Website: https://www.analyticsaholic.com/

LinkedIn: https://www.linkedin.com/in/raymondsondak/

CHAPTER 29

WHAT AI REVEALS ABOUT HOW WORK ACTUALLY HAPPENS

By Jennifer L. Stapleton
HR & Organizational Consultant, AI Strategist
Gold Beach, Oregon

Learning never exhausts the mind.
—Leonardo da Vinci

When people ask me about artificial intelligence, they often assume my interest began through my work in human resources and organizational consulting. In reality, my fascination with AI began years earlier through books, emerging technologies, and questions about the future relationship between people and intelligent systems. My career, however, has been built around people, organizations, and work. For more than two decades, I have worked in human resources, recruiting, workforce strategy, and organizational consulting, helping organizations hire talent, develop leaders, solve operational challenges, and navigate change.

In 2023, I found myself working with AI in a way I hadn't expected. As a human resources subject matter expert, I began helping train AI systems using my professional knowledge and experience. At the time, I viewed the work primarily as a technology project and assumed the most interesting lessons would be about artificial intelligence itself. What surprised me, however, was that many of the lessons had very little to do with technology. Instead, the experience began changing the way I thought about work.

Like many people, my early professional view of AI was shaped by questions about capability. Could it write content, analyze information, automate tasks, replace jobs, or increase productivity? These are understandable questions, and they remain at the center of many conversations about AI today. Yet after spending time both helping train AI and later incorporating it into my own work, I began noticing something else.

The most valuable insights were not necessarily about the technology itself. They were about the individuals, teams, and organizations using it. They were about how decisions were made, how information was shared, how work moved through organizations, and how individuals interacted with knowledge itself. In many ways, AI became a mirror. It reflected strengths that already existed, exposed inefficiencies that had been overlooked, highlighted assumptions people didn't realize they were making, and revealed patterns that had often been hidden in plain sight.

This is one of the reasons I believe AI represents far more than another technology trend. Throughout my career, I have watched organizations adapt to major disruptions. The internet changed how businesses communicate. Cloud technology changed how information is stored and accessed. Remote work changed where work happens. AI feels different, not because it is simply another tool, but because it has the potential to influence how people think, learn, solve problems, make decisions, and create value.

That shift may ultimately prove more significant than any individual software platform or technology application. It is also why I believe the organizations and individuals who thrive during this next disruption will be those who pay attention not only to what AI can do, but also to what AI is showing them.

More Than a Tool

Most conversations about artificial intelligence begin with capability. Some want to know what AI can do, how quickly it can perform a task, and whether it can save time or increase productivity. Organizations ask how AI can improve efficiency, reduce costs, automate work, or help employees accomplish more with fewer resources. These are reasonable questions and, in many cases, they are exactly what draw individuals and organizations to AI in the first place.

That assumption is understandable. Throughout modern history, organizations have adopted new technologies to improve communication, streamline operations, and increase access to information. From email and cloud computing to mobile devices and collaboration platforms, each innovation has helped people work in new and more effective ways. It is natural to assume AI simply represents the next step in that progression.

My own experience, however, has led me to view AI somewhat differently. While it can certainly automate tasks and generate results, I have found that much of its value emerges through interaction. One of the most surprising lessons I learned was that the value rarely comes from a single exchange. Early on, I expected AI to function much like a search engine: ask a question, receive an answer, and move on.

What I discovered was that the first answer was often just the beginning of the conversation. As I provided additional context, asked follow-up questions, or explored a different perspective, the discussion would frequently evolve in unexpected ways. Sometimes it challenged my assumptions, revealed connections I hadn't considered, or even made me laugh. What started as a simple request could turn into a much broader exploration of an idea, a problem, or a possibility.

Over time, I began to notice something else. The interaction felt less like retrieving information and more like engaging with a thinking partner. Not because the technology was replacing human judgment, creativity, or experience, but because it was helping me explore them more deeply. In some ways, the conversation began to feel like an extension of my own thinking: a place where ideas could be tested, refined, and expanded before being put into action.

The experience changed how I thought about AI. Rather than seeing it solely as a tool that delivers answers, I began seeing it as a resource that could help organize thinking, evaluate alternatives, and explore possibilities I might not have considered on my own. In many situations, the conversation became just as valuable as the answer itself because the quality of the outcome depended not only on the technology, but also on the engagement, curiosity, expertise, and critical thinking brought to the interaction.

That distinction is important because many individuals and organizations are still evaluating AI primarily through the lens of automation. They are asking which tasks can be completed faster or which processes can be delegated to technology. While those benefits are real, they represent only part of the opportunity. AI is also changing how we interact with information, approach learning, and think through complex problems. As experimentation continues, it is becoming clear that AI is not simply altering how work gets done. It is influencing how we approach the work itself.

Understanding that difference may be one of the most important lessons of the AI transformation. The future will not be shaped solely by the organizations that adopt AI the fastest. It will be shaped by those who learn how to work alongside it most effectively.

How AI Changes the Way We Think and Work

One of the reasons AI feels different from previous technologies is that it doesn't simply change what we do. It has the potential to change how we approach the work itself. For decades, knowledge workers have spent a significant portion of their time searching for information, organizing thoughts, conducting research, creating content, and solving problems. While technology has made many of these activities faster, the fundamental process remained largely unchanged. People gathered information, analyzed it, formed conclusions, and then took action. AI is beginning to alter that process.

Today, someone can explore a new idea, summarize a complex topic, compare multiple perspectives, draft a proposal, analyze data, or brainstorm solutions in a fraction of the time it may have taken in the past. The result is not simply greater speed. It is often a different relationship with information itself. Rather than spending hours collecting information, people can spend more time evaluating it. Rather than starting every project from a blank page, they can begin with a foundation and focus their energy on refinement and improvement. Rather than becoming overwhelmed by the volume of available information, they can use AI to organize, summarize, and identify patterns that might otherwise be difficult to see.

What interests me most about this shift is not the efficiency itself, but what becomes possible because of it. As AI reduces the time spent on routine tasks, it creates opportunities to focus on higher-value activities. Creativity, strategy, relationship building, innovation, and critical thinking become increasingly important. In many cases, the competitive advantage is no longer access to information. Information has become abundant. The advantage lies in judgment, interpretation, and the ability to apply knowledge effectively.

I have also observed that AI often encourages curiosity. Individuals who begin experimenting with it frequently discover new ways to approach problems, explore unfamiliar topics, and challenge assumptions. Questions that might have remained unasked suddenly become easier to investigate. Ideas that once required significant time and effort to explore become more accessible. In that sense, AI can become a catalyst for learning, not because it replaces expertise, but because it lowers the barriers to exploration.

At the same time, AI is not a substitute for thinking. If anything, it highlights the importance of thinking. AI can generate information, but it cannot determine what matters most. It can provide alternatives, but it cannot define goals. It can assist with analysis, but it cannot replace wisdom, experience, judgment, or human responsibility. This is where I believe many discussions about AI miss the mark. The conversation often focuses on what AI will replace rather than what it may enable. While some tasks will undoubtedly change, the larger opportunity may be the ability to redirect human time and energy toward activities that create greater value.

Throughout history, technological advances have reduced the effort required to perform routine work. AI has the potential to do the same for many knowledge-based activities. The question is not simply what people will do less of. The more important question is what people will be able to do more of. For individuals willing to embrace continuous learning, experimentation, and adaptation, AI offers an opportunity to expand their capabilities rather than diminish them. As AI continues to evolve, the most successful individuals may not be those who know the most about the technology itself. They may be those who learn how to combine human strengths with AI capabilities in ways that enhance both.

What Organizations Are Beginning to Discover

As individuals become more comfortable working with AI, organizations are beginning to ask a different set of questions. The conversation shifts from personal productivity to broader considerations about teams, workflows, decision-making, and organizational effectiveness. Many leaders initially approach AI as a technology initiative, evaluating software platforms, exploring use cases, establishing policies, and identifying opportunities for automation. These are important steps, but they are often only part of a much larger conversation. What I have found most interesting is that AI frequently reveals things that were already present within an organization. It doesn't necessarily create new challenges as much as it makes existing challenges more visible.

When AI is introduced into a process, organizations often discover that people perform the same work differently. Information may exist in multiple locations, decision-making authority may be unclear, responsibilities may overlap, and processes that appear straightforward on paper may be far more complex in practice. None of these situations are unusual. Organizations naturally evolve over time as new systems, processes, and workarounds are introduced to solve immediate challenges. Most of these changes are made with good intentions, but as they accumulate, they become part of the normal operating environment and are rarely questioned. As organizations attempt to automate, streamline, or accelerate work, they are often forced to examine how work actually happens. Questions that may have gone unasked suddenly become important. Who owns this process? Where does this information come from? Why are there multiple versions of the same document? How are decisions made? What happens when a key employee is unavailable?

For many readers, this may bring to mind the famous TPS reports scene from the movie *Office Space*. If you've ever worked in a large organization, there's a good chance you've had your own TPS report experience. What made the scene memorable wasn't the missing cover sheet. It was the fact that three different people felt compelled to remind Peter about the same issue. While intended as comedy, the scene captured something many organizations experience: processes, approvals, reports, and checkpoints that have accumulated over time until the purpose behind them is no longer clear.

The computer scientist and naval officer Grace Hopper once said, "The most dangerous phrase in the language is, 'We've always done it this way.'" The enduring relevance of that observation may help explain why AI is uncovering so many opportunities for improvement. When organizations begin examining how work actually happens, they often discover that long-standing processes are being followed out of habit rather than purpose.

Viewed through a modern lens, the scene feels surprisingly relevant. One of the first things AI tends to expose is work that exists primarily to support other work. As organizations explore opportunities to automate or streamline processes, they often find themselves asking questions that should have been asked years earlier. Why does this step exist? Who uses this information? What value does this report provide? Is there a simpler way to accomplish the same outcome? Or perhaps the most uncomfortable question of all:

If we were designing this process today, would we create it this way in the first place?

In many cases, the answers reveal opportunities for improvement that have little to do with AI itself. This is one reason I believe the most successful AI initiatives are rarely technology projects alone. They are also conversations about people, communication, decision-making, accountability, and the flow of work throughout the organization. The technology may serve as the catalyst, but the lessons often extend far beyond the technology itself. Organizations that recognize this tend to approach AI differently. Rather than viewing it as a replacement for human contribution, they begin viewing it as an opportunity to better understand how their teams operate and where greater effectiveness can be achieved. The goal becomes not only working faster, but working smarter and more intentionally.

As AI adoption continues to grow, I believe some of the greatest benefits will come from this increased visibility. The organizations that thrive may not be those with the most advanced technology. They may be the organizations that are willing to learn from what the technology reveals and use those insights to strengthen the way people, processes, and information work together.

Why This Disruption Feels Different

Throughout my career, I have watched organizations adapt to significant change. New technologies arrive, new processes emerge, and people gradually learn new ways of working. While each disruption creates uncertainty, most eventually become part of everyday life.

AI feels different, not simply because it can perform tasks more quickly or process large amounts of information, but because it is beginning to influence activities that many people have traditionally associated with human thinking. It can help generate ideas, analyze information, explore alternatives, organize knowledge, and accelerate learning. As a result, the conversation extends far beyond technology. It now includes creativity, decision-making, communication, problem-solving, and the future of work itself.

That reality creates understandable excitement as well as understandable concern. Questions about workforce skills, productivity, organizational effectiveness, and the future of work continue to dominate headlines and boardroom discussions. While no one can predict exactly how AI will evolve, I believe the more important question is how people will evolve alongside it.

Throughout history, adaptability has often proven more valuable than certainty. The individuals and organizations most likely to thrive will not necessarily be those who adopt every new technology first. They will be those who remain curious, continue learning, and develop the ability to work effectively alongside new tools and new ways of thinking.

Perhaps that is what makes this disruption different. AI is not simply changing how work gets done. It is encouraging people to reconsider how they learn, how they solve problems, and how they create value. Those who embrace that opportunity may discover that the greatest transformation is not technological at all.

Conclusion

When I first began working with AI, I expected to learn about technology. What I did not expect was how much it would change the way I thought about work, learning, and problem-solving. Over the past several years, I have watched conversations about AI evolve from curiosity and experimentation to discussions about transformation, opportunity, and disruption. Along the way, I have come to believe that many of the most important lessons are not about artificial intelligence itself. They are about how individuals and organizations learn, adapt, make decisions, and create value in a rapidly changing world.

AI is often discussed as though it exists separate from us, as though it is something arriving from the outside that will either solve our problems or create new ones. My experience has been far more nuanced. AI reflects the questions we ask, the goals we pursue, and the intentions we bring to its use. It can help us move faster, but it can also encourage us to think more deeply. It can automate routine tasks, but it can also create space for creativity, learning, and innovation.

Perhaps that is why I remain optimistic. Not because I believe AI will have all the answers, but because I believe people will continue asking better questions.

The organizations that thrive in the years ahead will not be defined solely by the technology they adopt. They will be defined by their willingness to learn, adapt, and rethink how work is done. Likewise, individuals who remain curious and open to new possibilities will be better positioned to navigate whatever changes lie ahead.

The AI transformation is often described as a technological revolution. It may ultimately prove to be something more. It may become an opportunity to reconsider how we learn, how we collaborate, how we create value, and how we work together.

And perhaps the most important question is not what AI will become. Perhaps the more important question is what we will become as we learn to work alongside it.

About the Author

Jennifer L. Stapleton is an HR and organizational consultant, business strategist, and entrepreneur with more than two decades of experience helping organizations navigate workforce challenges, leadership development, organizational effectiveness, and business transformation. A Colorado native now living on the southern Oregon coast, Jennifer's career has spanned industries including technology, healthcare, government contracting, retail, supply chain, and energy.

In 2023, she began contributing as a human resources subject matter expert in the development and training of artificial intelligence systems. What began as an opportunity to apply her professional expertise evolved into a deeper exploration of the relationship between people, technology, and the future of work. Today, Jennifer combines her experience in human resources, organizational consulting, and

AI to help leaders and organizations adapt to change, solve complex challenges, and create more effective ways of working.

Jennifer is the founder of JLS Worldwide Inc. and owner of The Blue House, a developing business and community space in Gold Beach, Oregon.

Email: Jennifer@jlsworldwide.com

Website: jlsworldwide.com

LinkedIn: https://www.linkedin.com/in/jennifer-stapleton-29b8b13/

CHAPTER 30

CURIOSITY AND COURAGE—THE STRATEGIC ENGINE OF AI TRANSFORMATION

By Ivo Strohhammer
Expert for Digital Transformation
Zug, Switzerland

We keep moving forward, opening new doors, and doing new things, because we're curious and curiosity keeps leading us down new paths.

—Walk Disney

For roughly $20 a month, the world was promised a revolution. That is the approximate cost of a premium AI subscription in 2026, less than a business lunch, or a rounding error in a multinational's quarterly travel budget. In boardrooms ranging from global corporations to agile SMEs, a pervasive and dangerous myth has taken root: that the "AI Transformation" is a software update you can simply buy, install, and ignore.

But here is the uncomfortable truth, and it cuts both ways. The subscription is the cheapest line item; the real cost lives in integration, training, redesigned processes, outside advisors, and sometimes new roles, a bill that falls hardest on smaller firms. Yet even a fully funded budget guarantees nothing. If your AI strategy starts and ends with a check, whatever its size, you are merely automating your current inefficiencies at greater expense.

I've seen this pattern repeat across industries. Organizations often fall into what I call the "status quo trap." They treat AI as a digital bandage for analog wounds, hoping that a chatbot can mask the smell of decaying processes. But AI is not a bandage; it is an accelerant. It will accelerate your success if your foundation is strong, but it will accelerate your collapse if your foundation is built on "the way we've always done it."

Why Do We Need to Transform?

The inescapable reality is that the "old way" is becoming a liability. In an age of exponential technology, the speed of innovation is accelerating at a rate that defies traditional planning cycles. Standing still is no longer a neutral position; it is the equivalent of moving backward. We transform because the competitive landscape has shifted, but also because the window for adaptation is narrowing.

This urgency is particularly acute in small and medium-sized enterprises (SMEs) that do not have the luxury of excess capacity. In these leaner environments, every hour spent on manual, repetitive "robotic" tasks is an hour stolen from strategic survival. We have to see the urgency to transform now, not as a future luxury, but as a current necessity. Transformation is the key to unlocking the dormant human potential that is currently being throttled by these outdated processes, allowing us to pivot before the disruption becomes a displacement.

There is a quieter reason the old way has become a liability: standing still no longer even delivers the safety it promises. In a market where competitors compound small, AI-driven gains week after week, the "safe" choice is quietly the most dangerous one. The threat is rarely one dramatic disruption; it is the slow erosion of relevance while the world moves on around you.

The Three-Dimensional Equation

To understand why so many AI initiatives stall, we must look at the fundamental equation of digital transformation:

DT = Technology x Process x People

Most corporate leaders are comfortable with the first two variables. They understand technology procurement and they are experts at iterative process improvement. But in the rush for "cheap and fast" ROI, the most volatile and valuable variable—people—is being dangerously neglected. And note that the equation multiplies; it does not add. You can buy the best technology and engineer flawless processes, but multiply them by a workforce that is fearful, untrained, or unconvinced, a people factor near zero, and the entire product collapses toward zero. This is the mathematics behind every stalled transformation.

We often see "transformation" projects where training is treated as an afterthought, a quick webinar scheduled a week before go-live. This isn't adoption; it's compliance. And it is not only people who get the afterthought treatment. Notice the trap hiding in that second variable: Leaders are masters of iterative process improvement, making the existing path faster and cheaper, but true transformation demands radical process redesign, the harder act of asking whether a given process should exist at all. It demands a radical rethink not just of what we do, but of who we are in the workflow.

Business Strategy vs. Grassroots Movements

While a top-down strategy provides the essential business direction, in today's world, it requires courage from leadership to push into undiscovered territory. It is no longer sufficient for management to simply steer the ship; they must have the bravery to venture into unmapped waters, letting go of traditional KPIs to support radical shifts.

However, this top-down courage is only half of the story. The bottom-up movement provides the soul of the transformation, it is the engine for validation, experimentation, and true adoption. This grassroots fire doesn't ignite by accident. It requires a workforce driven by a clear purpose and a relentless curiosity mindset. This movement can only thrive within a general culture of experimentation that embraces failure as a necessary byproduct of discovery. For a transformation to be successful and sustainable, the people on the front lines must understand not just *how* to use the tools, but *why* they should care. Adoption doesn't happen because of a CEO mandate; it happens when curiosity is protected and individuals find the personal incentives to innovate.

But curiosity here means far more than enthusiasm or play. At its core it is a discipline of critical thinking, the relentless habit of asking why. Why does this process exist? Why do we do it this way, and does it still earn its place? Curiosity is what exposes the broken and the obsolete; courage is what changes them. The two are inseparable: understanding without the nerve to act changes nothing, while acting without understanding only automates the wrong thing faster. And the why can only be asked aloud where people are genuinely free to speak up, which is why challenging the status quo is, in the end, as much a question of culture as of character.

Transforming Without a Map

If the case for transforming is clear, the harder question is the one that keeps leaders awake at night: Transform toward what? We are asked to commit budget, time, and credibility to a future no one can actually see. The technology landscape reshapes itself every few months, and today's indispensable platform is tomorrow's footnote. Acting decisively is essential, but acting in the wrong direction can cost as much as standing still. This is the real anxiety beneath most stalled initiatives: not a lack of will, but a lack of certainty about where to place the bet.

The instinct is to wait for clarity, but in an exponential environment it never fully arrives: By the time a technology is "proven", the advantage has gone to whoever moved while it was still ambiguous. Transformation is not picking the right answer from a list; it is moving forward intelligently while the answer is unknown.

Betting Without Certainty

Once you accept that no bet is risk-free, the goal shifts from making the perfect choice to making smart, survivable ones. A few principles help:

- **Bet on capabilities, not vendors:** Tools change; the underlying capability you are building, automating a workflow, augmenting a decision, freeing human time, endures.
- **Keep your bets reversible:** Favor experiments you can unwind cheaply over grand, irreversible commitments.
- **Protect what is truly yours:** Technologies are rented; your real assets are your data, your domain knowledge, and your people's ability to learn.
- **Make many small bets, not one large one:** You cannot predict which experiment will pay off, so run several.
- **Decide on a cadence, not forever:** In a stable world you decide once and move on; in this one, every bet deserves a review date.

This is where courage and curiosity stop being abstractions and become a working method. You do not need a map of the future; you need a compass, a clear sense of the outcomes you are pursuing, and the willingness to take the next honest step, learn from it, and correct course. In a world that refuses to hold still, navigating without certainty is not an inconvenience to be solved. It is the core competency of the age.

What's In It for the People?

To move people from fear to curiosity, we must answer the most critical question: What is in it for me?

- **More Impact, Less Busy-Work:** AI allows individuals to shed the soul-crushing, repetitive tasks that fill the forty-hour week. It moves the worker from "doing" to "orchestrating," allowing them to focus on high-value problems that require human judgment, empathy, and creativity.
- **Self-Fulfilling Roles:** Instead of being a cog in a machine, the individual becomes the architect of their own digital team, in work aligned with their strengths.
- **The Survival Edge:** Those who lead the transformation aren't just surviving the disruption; they are defining it. By becoming an Orchestrator early, an individual increases their market value and secures their place in the future of work.

Clear incentives turn change from a threat into an opportunity, and a corporate project into a movement

From Task-Doer to Hybrid Orchestrator

This is not just another IT project; it is a fundamental shift in human identity, a transition from being a "task-doer" to leading a hybrid workforce.

In this new reality, everyone is becoming a manager now.

New roles are emerging as we move into hybrid teams of human colleagues and digital workers. Whether you direct ten people or a dozen autonomous AI agents, the skill is the same: directing outcomes rather than executing tasks. This is not a claim that everyone suddenly wants to manage people or that steering an agent equals leading a human. The point is harder to escape: The center of gravity of valuable work is shifting from doing the task to directing the outcome, and that is a leadership skill, whether the team is human, digital, or both.

This is the era of the hybrid orchestrator.

The barriers to high-impact innovation have collapsed. An individual can now achieve what once required a large team of experts and a multi-million-dollar budget. Reaching that level demands a cultural shift, a "dual-engine" model: a top-down vision that grants the permission to experiment, paired with a bottom-up curiosity movement that decides what we build.

Building a Culture of Experimentation

To build the "dual-engine" model I have described, an organization must solve for the most significant bottleneck in any digital transformation: the fear of failure. If curiosity is the fuel of the AI era, then a safe space is the engine in which it burns. Without a culture that explicitly permits, and even rewards, experimentation, the most advanced AI tools will simply sit on the digital shelf. Transformation requires an environment where innovation is not crippled by legacy processes or the looming threat of personal obsolescence.

Psychological Safety as a Technical Requirement

Psychological safety is not a soft skill; it is a technical requirement. If an employee feels that a "hallucinating" AI or a failed automation script will lead to a negative performance review, they will never push the boundaries. They will stick to the "safe," manual, and inefficient methods they know. Leaders must provide a safe space for innovation, an environment where the focus shifts from "Who made this mistake?" to "What did we learn from this iteration?"

The Automation Paradox: Fear of Job Loss

One of the most silent yet destructive forces in AI transformation is the fear of self-cannibalization: the belief that automating your own tasks means writing your own pink slip. It would be dishonest to call that fear groundless. Some roles will shrink and some will disappear, and a transformation that creates no redundancy at all is usually just theatre.

The answer is not a hollow guarantee that every job is safe, which fools no one and only feeds the cynicism that kills curiosity. It is an honest bargain: Those who learn to orchestrate rather than execute are first in line for the higher-value roles the transformation creates, provided that promise is backed by real retraining and not just rhetoric. Innovation then stops being a threat to survival and becomes the most reliable route to it.

Creating Capacity and Removing Crippling Processes

You cannot be curious if you are 100% utilized. This is the "luxury of capacity" that many SMEs struggle with. If every minute of a forty-hour week is booked with "execution," there is zero space for "exploration."

Strategic leaders must purposefully build "slack time", such as capping utilization targets at 85% or mandating "innovation Friday afternoons", into the system. This isn't wasted time; it is the research and development budget of the human mind. Furthermore, we must dismantle the crippling processes, the endless layers of approval and the "innovation-by-committee" approach, that stifle the speed of curiosity. AI moves at the speed of thought; if your internal approval process moves at the speed of paper, the transformation will die in the inbox.

Capacity, though, is not only a matter of hours; it is also a matter of energy. Curiosity and courage are renewable, but they are not infinite, and a workforce asked to reinvent itself every week without pause will not stay curious for long. It will burn out and retreat to the safety of the old way. Sustained transformation, therefore, has a human metabolism: Leaders must pace the change, protect time to absorb and recover, and treat adaptation as a marathon run at a steady stride rather than a sprint repeated until exhaustion. The goal is not maximum change at maximum speed, but a rhythm the organization can actually keep.

A true failure culture provides the tools, the time, and the psychological safety to try, fail, learn, and iterate. It is the only way to move from a culture of compliance to a culture of orchestration.

Thriving Through the Disruption: Actionable Success Advice

Surviving this disruption is not a matter of luck; it is a matter of strategic intent. Whether you are leading a multinational or building your own career, the following actionable steps will help you move from a "task-doer" to a "hybrid orchestrator."

For the Individual: Your Innovation Journey

1. **Reclaim your "childish" curiosity:** Our school systems often train curiosity away, replacing it with a search for the "right" answer. In the AI era, there is no single right answer, only better iterations. Start by asking "What if?" at least once a day regarding a manual task you perform.
2. **Identify the "cool" tasks:** Don't just automate the boring things; use AI to find the tasks that give you a "superpower." If you can achieve an enterprise-level result with a digital agent, you aren't just an employee; you are a force multiplier.
3. **Build your own "digital workforce":** Start treating AI agents not as search engines, but as digital employees. Give them personas, specific job descriptions, such as "You are an aggressive proofreader hunting for logical fallacies",

and 360-degree feedback. If you don't learn how to manage digital workers now, you won't be ready to lead human ones tomorrow.

For the Corporate Leader: Building the Sandbox

1. **Implement the "pragmatic post-mortem":** Move away from blame. When an AI project fails, hold a ritual focused on three things: root-cause analysis, prevention strategies, and, most importantly, sharing the learning globally. Make the "freedom to fail" a visible, rewarded behavior.
2. **Create "AI guilds":** Consultants often miss the real use cases that exist in the "domain knowledge" of your employees. Empower grassroots "AI guilds", communities of practice where business people (not just IT) share their experiments and successes.
3. **Guarantee the "elevation promise":** Explicitly communicate that anyone who "automates themselves out of a task" is automatically a candidate for higher-level strategic roles. Remove the fear of obsolescence and replace it with the promise of elevation.

Reclaiming the Human Superpower

The AI transformation is not coming; it is already here. The "status quo trap" is comfortable, but it is terminal. To thrive, we must recognize that while the tools are digital, the engine is human.

By combining the courage of top-down leadership with the curiosity of a bottom-up movement, we can create a dual-engine culture that doesn't just survive the disruption but leads it. The future belongs to the hybrid orchestrators, those who have the courage to dismantle the old ways and the curiosity to imagine what's next.

Now is the time to start your innovation journey. Don't wait for a $20 subscription to change your life. Reclaim your curiosity, build your courage, and start orchestrating your future today.

About the Author

Ivo Strohhammer is a strategist and digital transformation architect operating at the precise intersection of technology, process optimization, and human leadership. His career is defined by a distinct "dual perspective" on change management: He has successfully navigated and driven internal innovation within the deeply structured, complex procedural landscapes of global powerhouses like Siemens and EY, while simultaneously advising small and medium-sized enterprises (SMEs) on how to unlock rapid gains in agility, quality, and market responsiveness.

Having engineered cross-border initiatives throughout his international career, Ivo has specialized in a challenge that trips up most modern organizations: overcoming cultural inertia and change resistance. He bridges the gap between sophisticated corporate strategy and pragmatic, boots-on-the-ground execution, always keeping the human factor at the center of digital shifts.

A passionate advocate for AI as the ultimate democratization of technology, Ivo is currently building an AI ecosystem and network to foster adoption and innovation in Switzerland. Ivo is also the author of the forward-looking *What If* book series, short stories about the future and the role humans and technology might play. His mission is to dismantle the "status quo trap" and turn everyday task-executors into leaders of a unified human-digital future.

Email: ivo.strohhammer@tosummit.ch

Website: www.tosummit.ch

LinkedIn: https://www.linkedin.com/in/ivo-strohhammer-07358972/

CHAPTER 31

WILL AI AUGMENT US OR MAKE US INCAPABLE OF CREATIVE THINKING?

By Sakina Syed
Microsoft IT Professional, AI Engineer and Consultant
Toronto, Canada

We shape our tools, and thereafter our tools shape us.
—Marshall McLuhan

Imagine sitting down to write, solve, or create, and instead of staring at a blank page, you open an AI tool. Within seconds, ideas begin to flow. Sentences form. Structures emerge. Possibilities expand faster than your mind can usually process.

You feel empowered. But also...uneasy. A quiet question lingers beneath the surface: "If AI can do this so easily, what happens to my ability to think?"

This tension defines our relationship with artificial intelligence today. On one hand, AI presents an unprecedented opportunity to enhance human creativity and productivity. On the other, it raises a deeply human fear: that we may be outsourcing not just effort, but imagination itself. That something subtle but essential may begin to fade the more we rely on it.

This concern is not new. History shows us that every transformative tool has sparked similar anxiety. When the printing press emerged, critics warned it would weaken memory and erode intellectual discipline. When calculators entered classrooms, many feared students would lose basic mathematical ability. When the internet became widespread, it was accused of fragmenting attention and creating shallow thinkers.

And yet, each of these innovations ultimately expanded human capability rather than diminishing it. They shifted how we think, not whether we think.

AI is no different, but it is more powerful, more intimate, and more cognitive than any tool before it. It does not just store knowledge. It participates in reasoning. It does not simply assist physical tasks. It engages directly with thought.

So the real question is no longer: Will AI make us incapable of creative thinking? The real question is: Who do we become when thinking itself can be augmented?

Because AI does not eliminate creativity, it reshapes it. And whether that reshaping becomes expansion or erosion depends entirely on us, our awareness, our habits, and our willingness to remain intellectually awake.

Redefining Creativity in the AI Era

To understand AI's impact, we must first rethink what we mean by creativity.

We often romanticize creativity as the ability to produce something entirely original, as if great ideas emerge in isolation from a single brilliant mind. But creativity has never worked that way. It has always been a process of connection: linking existing ideas in new ways, reframing problems, and applying knowledge across contexts.

Every breakthrough—whether in art, science, or business—builds upon something that came before. Even the most original thinkers are, in truth, master integrators. No thought exists in a vacuum. Every insight is shaped by what we have read, experienced, observed, and internalized over time.

In that sense, AI is not introducing something unnatural into the creative process. It is simply making this process more visible and more scalable.

AI systems are trained on vast bodies of human knowledge and expression. When we interact with them, we are not engaging with an independent creator, but with a system that reflects patterns of human thought at scale, synthesizing the past to generate possibilities in the present. And that changes our role.

Creativity is no longer defined purely by generating content from scratch. Instead, it shifts toward curation, direction, and refinement. We move from being sole originators to becoming editors of possibility, deciding what matters, what resonates, and what deserves to be developed further.

At the heart of this shift lies a critical distinction:

- Outsourced thinking: relying on AI outputs without engagement
- Augmented creativity: using AI to expand, challenge, and deepen our thinking

The difference is not in the tool. It is in the intention behind its use. We are no longer just creators. We become:

- Curators of possibility
- Directors of ideas
- Editors of meaning

The shift is subtle, but profound. Creativity moves from generation to judgment. It is no longer defined by how much we can produce, but by how well we can discern, what to keep, what to refine, and what to elevate.

AI does not remove the need for creativity. It raises the standard for it. It demands greater clarity, stronger judgment, and deeper perspective. And in doing so, it does not make creativity obsolete. It makes shallow creativity obsolete and calls forth something far more intentional in its place.

The Augmentation Advantage

When used intentionally, AI becomes one of the most powerful amplifiers of human creativity ever created. One of its greatest strengths is its ability to eliminate the friction that often blocks creative thinking. The blank page, long considered the enemy of creativity, loses its power when ideas can be generated instantly. The inertia that stops us from starting begins to dissolve.

Writers can explore multiple angles in minutes. Designers can visualize dozens of variations in seconds. Professionals can simulate strategies before committing to one path. What once took hours of effort now takes moments of exploration. This speed does not replace thinking, it expands it.

AI enables what we might call "cognitive leverage": the ability to do more with the same mental effort. Instead of spending energy generating raw material, we can focus on shaping, interpreting, and improving it. For example:

- A writer can generate multiple introductions, then refine the one that aligns best with their intent and voice.
- A strategist can test different scenarios and focus on evaluating outcomes rather than constructing them from scratch.
- A student can explore multiple explanations of the same concept, deepening understanding through comparison.

In each case, AI handles the initial expansion of ideas, allowing humans to engage in higher-order thinking: analysis, synthesis, and judgment.

Rather than replacing creativity, AI redistributes where creativity happens. It shifts it from effort-intensive production to insight-driven refinement. And in doing so, it has the potential to elevate not just what we create, but how we think. From "What can I produce?" to "What is worth keeping?" The result is not less creativity. It is higher-order creativity.

The Real Risk: Passive Thinking

While AI offers tremendous potential, there is a legitimate concern, and it deserves to be taken seriously: The risk is not that AI will make us incapable of thinking. The risk is that we may choose not to think.

When AI is used passively, copying outputs without questioning, accepting answers without reflection, it can lead to a subtle erosion of critical thinking. Over time, this can create a habit of intellectual disengagement, where convenience replaces curiosity.

This is not a new phenomenon. Technologies have always reshaped cognitive habits. GPS has reduced our reliance on spatial memory. Calculators have reduced mental arithmetic. Search engines have changed how we retain and retrieve information.

AI, however, operates at a deeper level. It engages directly with reasoning, writing, and problem-solving. It does not just provide information; it structures it, interprets it, and presents it as if it were thinking.

This makes the distinction between passive use and active use more important than ever.

- Passive Use: Accepting outputs as final, using AI as a shortcut (cognitive stagnation)
- Active Use: Questioning, refining, and challenging outputs (cognitive expansion)

The difference determines the outcome. If AI becomes a substitute for thinking, it can lead to cognitive stagnation. But if AI becomes a partner in thinking, it can lead to cognitive expansion. The tool does not decide the outcome. The user does. Therefore, the tool is not the problem, disengagement is.

The Stigma of Using AI

Despite its growing presence, AI usage still carries a subtle stigma. Many people feel hesitant to admit they use AI tools. There is an underlying perception that relying on AI is a form of "cheating". That it shortcuts effort and reduces authenticity.

This stigma is rooted in long-standing cultural values. We tend to equate effort with worth. If something is produced quickly or efficiently, we assume it must be less meaningful. But this perspective overlooks a key reality: Throughout history, progress has always been driven by tools that make tasks easier and expand human capability. Calculators did not diminish mathematicians; they allowed them to solve more complex problems. Spell check did not diminish writers; it allowed them to focus on expression rather than mechanics. The internet did not diminish knowledge; it expanded access to it. AI is part of this same continuum.

The irony is that while society rewards outcomes, it often judges the methods used to achieve them. Yet, in practice, what matters is the quality, insight, and impact of the output, not how labor-intensive the process was.

Using AI effectively is not a shortcut, it is a skill. It requires:

- Asking thoughtful, precise questions
- Interpreting responses critically
- Applying personal judgment and context

The stigma surrounding AI will fade, just as it has with every transformative tool. But for that to happen, we must shift the conversation from "Are you using AI?" to "How well are you using it?" Because the future will not belong to those who avoid AI. It will belong to those who use it with intention and intelligence.

The Blip: Is AI Causing Us to Forget How to Think or Remember?

There is a growing, almost imperceptible shift happening in how we remember. Think about the last time you needed to recall something: a fact, a date, a concept, or even an idea. Did you pause and try to retrieve it from memory? Or did you instinctively reach for a device, type a prompt, and retrieve the answer instantly?

That moment, small, automatic, and nearly invisible, is what we can call "the blip." The blip is not dramatic. It is not a collapse of intelligence or a sudden loss of memory. It is subtle. Quiet. Incremental. It is the repeated decision to retrieve rather than recall, to access rather than engage, to resolve uncertainty instantly rather than sit with it.

Over time, these moments accumulate, and they begin to reshape how we think. But is this truly a loss? Or is it a transformation?

Humans have always externalized memory. Writing, books, libraries, and search engines extended our ability to store and retrieve knowledge beyond the limits of the human brain. AI is simply the next evolution.

The difference is that AI doesn't just store information. It contextualizes it, interprets it, and delivers it with fluency. This creates the impression that remembering is no longer necessary at all. And in some ways, it isn't. But what is changing is not our capacity to remember, it is what we choose to remember. And so, the question changes: Not "Are we forgetting?" But "What are we choosing not to remember anymore?"

There is, however, a trade-off. When we consistently offload memory, we may weaken our habit of mental effort. The discipline of thinking deeply, wrestling with ideas, and forming independent conclusions can begin to fade if every answer is instantly available.

The blip, then, is not about forgetting. It is about choosing not to engage before retrieving. Yet, this shift can also be reframed.

In an AI-enabled world, memory becomes less about storing isolated facts, and more about understanding connections, patterns, and meaning. We begin to remember frameworks rather than fragments. Relationships rather than details. Principles rather than specifics. This aligns with higher-order thinking. Instead of memorizing everything, we learn how to think about anything. The key is balance.

We must resist the reflex to immediately outsource every question. Some friction is necessary. It is in the pause in the effort to recall, question, and interpret that thinking is strengthened.

To avoid the blip trap:

- Pause before prompting.
- Think first, then verify.
- Question outputs, don't just consume them.
- Practice independent thinking regularly.

When used consciously, AI does not weaken memory—it reshapes it.

The real question is no longer: How much can we remember? But rather:

What is worth remembering in a world where everything is accessible? Because the future will belong not to those who remember the most but to those who can think deeply with what they remember and what they can access.

The Question of Originality: What Is Still "Yours"?

As AI becomes more integrated into the creative process, a quieter, more personal question begins to emerge: If AI helped create it, is it still mine? This question is not about technology; it is about identity.

For centuries, creativity has been closely tied to authorship. We define ourselves through what we produce, believing that our ideas, our words, and our creations are direct expressions of who we are. Ownership, in this sense, feels inseparable from originality. But in an AI-enabled world, the boundaries begin to blur.

If an idea is sparked by an AI suggestion, refined by your judgment, and shaped by your perspective, where does authorship truly begin? The answer is not as binary as it once seemed. Because ownership has never been about creating in isolation. It has always been about interpretation, intention, and voice.

Two people can start with the same input and create entirely different outcomes. Not because the material is different, but because *they are*. Your lived experiences, your values, your intuition, and your judgment—all of these shape how you select, refine, and elevate ideas. AI may expand the field of possibilities, but it cannot determine which ones matter. That remains deeply human. In this sense, originality is evolving.

It is no longer defined solely by where an idea comes from, but by how deeply it is transformed. AI can assist in generation. But meaning, true meaning, still requires ownership. And ownership is not about doing everything alone. It is about being able to say, "This reflects how I think. This represents what I believe. This is mine."

In the age of AI, authenticity does not disappear. It becomes *more* intentional.

From Tool to Thought Partner

To fully realize AI's potential, we must change how we see it. Most tools extend physical capability. AI extends cognitive capability. This makes it more than just a tool, it becomes a thought partner. When we treat AI as a partner, our behavior shifts. Instead of asking for answers, we engage in dialogue. We explore possibilities, test assumptions, and refine ideas through iteration. This introduces a new discipline: prompting is thinking.

The way we ask questions determines the quality of insight we receive. Poorly framed prompts produce surface-level results. Thoughtful prompts unlock deeper, more nuanced responses.

In this way, AI reflects our thinking back to us. It forces clarity. It demands structure. It highlights gaps in reasoning. Rather than reducing thinking, AI makes it more visible and more deliberate. Used well, it does not just provide answers. It makes us better thinkers.

Rewiring Ourselves for the AI Era

Adapting to AI requires more than learning tools. It requires rewiring how we approach thinking itself.

1. From Knowing to Navigating

Knowledge is no longer scarce. It is abundant and accessible. The advantage shifts from memorization to navigation: the ability to ask the right questions, find relevant insights, and connect ideas meaningfully.

2. From Creation to Direction

Creativity becomes less about producing from nothing and more about guiding, shaping, and refining ideas. Humans become orchestrators rather than sole creators.

3. From Individual to Hybrid Intelligence

The future is not human or AI, it is human and AI. The most effective thinkers will combine intuition, judgment, and lived experience with AI's speed and scale.

4. Building New Habits

- Think before prompting.
- Challenge what you receive.
- Add personal perspective.
- Alternate between independent and AI-assisted work.

Rewiring is intentional. But those who do it will not become less capable. They will become more powerful thinkers.

A Framework for Healthy AI Use

To use AI effectively, we need structure:

- Generate: Explore ideas.
- Interrogate: Question outputs.
- Refine: Add meaning.
- Own: Make it yours.

Each step matters. Skipping interrogation leads to blind acceptance. Skipping refinement limits depth. Skipping ownership disconnects the work from your voice.

When practiced consistently, this framework transforms AI from a passive tool into an active thinking environment, one that sharpens judgment rather than dulling it. It creates a rhythm of engagement where you are constantly interpreting, shaping, and deciding. This turns AI into a thinking partner not a crutch.

The Future of Creativity

AI will not reduce creativity; it will expand it and democratize it. More people will create. More ideas will be expressed. Barriers will fall. But what stands out will change. Not volume but depth. Not speed but perspective. Not output but judgment.

As access increases, differentiation will no longer come from producing more. It will come from thinking better. The ability to interpret, connect, and assign meaning will become the true creative advantage. AI can generate ideas. But only humans can assign meaning. Creativity will not become less human. It will become more deeply human than ever before.

The Choice Is Ours

So, will AI augment us, or make us incapable of creative thinking?

The answer is not predetermined. AI does not decide how we think. We do. We can engage deeply, or disengage entirely. We can use AI to expand our minds, or to bypass them. We can treat it as a shortcut, or as a catalyst for deeper insight. Each interaction becomes a choice: to question or to accept, to reflect, or to move on.

The technology is neutral. The outcome is human. In the end, AI will not make us less creative. It will reveal how willing we are to think, how committed we are to staying intellectually present in a world designed for speed and ease. Because in the age of AI, creativity is no longer defined by what you can produce alone, but by how powerfully you can think, question, and create with it, and how intentionally you choose not to rely on it when your own thinking matters most.

About the Author

Sakina Syed is a dynamic senior data and AI consultant, AI engineer, and enthusiast with a stellar track record in the computer software industry, including notable tenures at tech giants like Microsoft. As a Microsoft AI engineer and Azure-certified professional, she excels in AI deployment, sales, management, teamwork, and leadership.

With a Bachelor of Science in Neuroscience and Mental Health Studies from the University of Toronto, Sakina has also enriched her expertise with courses in artificial intelligence, business management, and IT. Her passion for writing and traveling adds a unique dimension to her professional persona.

Sakina's enthusiasm for business and the application of business psychology to understand consumer needs is matched by her extensive seven-year experience in customer service. She is adept at interpreting consumer feedback and statistical data, enabling her to identify market requirements with precision. Her strong passion for team management and leadership drives her dedication to delivering exceptional customer experiences and fostering business success.

Email: sakina@youraiconsultant.ca

Website: www.youraiconsultant.ca

LinkedIn: https://www.linkedin.com/in/sakina-syed/

CHAPTER 32

SELF-DETERMINATION AI: THE ENABLING PHILOSOPHY BEHIND AI TRANSFORMATION AMID DISRUPTION

By Gary Vause, II
Professor, CEO, Vause Computer Systems, LLC
Atlanta, Georgia

My journey toward what I now call "self-determination AI" did not begin with artificial intelligence. It began long before the current wave of AI innovation, in the fields, classrooms, military bases, businesses, and communities that shaped my understanding of ownership, learning, and value creation.

I often describe my personal story as "from fruit picker to professor." Growing up in Central Florida, I learned early that opportunity was rarely handed to you. It was cultivated. My great-grandparents were poor Mississippi farmers who understood that prosperity came from transforming available resources into something of value. That lesson followed me through my service in the United States Air Force during the Cold War, my graduate studies in the United Kingdom, my professional career with KPMG, my years in higher education, and ultimately my research into digital transformation and artificial intelligence.

Every major technological revolution forces society to confront a deeper question about what it means to be human. Steam power redefined labor. Electricity redefined industry. The internet redefined communication. Artificial intelligence is redefining something far more personal: our relationship to capability itself.

For the first time, we are surrounded by tools that can think, generate, predict, and automate. This shift has created both excitement and unease. People sense the magnitude of the moment, yet many feel unprepared for what comes next. They worry about relevance, opportunity, and whether they will have a meaningful place in a world increasingly shaped by intelligent systems.

Here is the truth at the center of *The AI Transformation: Thriving Within Civilization's Next Big Disruption*: Technology alone does not determine who thrives. Human agency does. The individuals and organizations that flourish in this new era will not be the ones with the most advanced tools. They will be the ones who understand how to use AI intentionally, strategically, and in alignment with their own values and goals.

This is why self-determination AI sits at the heart of the transformation. It is not a technical framework. It is a human one. It provides the mindset, structure, and psychological foundation required to turn AI from a source of uncertainty into a source of empowerment.

Self-determination AI reframes the central question of the era. Instead of asking what AI can do, it asks what people can become when they learn to orchestrate AI with ownership and purpose. This chapter explores how that philosophy works and why it is the enabling factor behind thriving in the disruption ahead.

The Shift from Tools to Partners

Throughout history, tools have extended our physical abilities. The wheel made it easier to move goods. The plow made it easier to cultivate land. The engine made it easier to travel and build. AI extends far beyond something far more consequential: our cognitive abilities.

This shift changes the nature of work in profound ways. When a machine can draft a report, summarize a meeting, or analyze a dataset, the value of human contribution shifts from execution to interpretation. The question is no longer whether you can perform the task, but whether you can understand what the task means.

AI can generate options, yet people must choose the direction.

AI can surface patterns, yet people must decide which patterns matter.

AI can accelerate thinking, yet people must determine the purpose behind the thought.

This distinction is not philosophical. It is strategic. Organizations that treat AI as a replacement for human capability will optimize efficiency but sacrifice adaptability. Individuals who rely on AI to think for them will become dependent rather than empowered.

Self-determination AI begins with a simple principle. AI can support thinking, but only people can provide understanding. Understanding is where leadership lives. It is where strategy is shaped. It is where meaning is created. And it is where human beings remain irreplaceable.

The Fruit Picker and the Digital Garden

This concept of the digital garden is deeply personal. Throughout my career, I have observed that many people work tirelessly to create value within systems they do not own. My research asks a simple but important question: How do we help people move from being consumers of technology to owners of digital assets, systems, knowledge, and outcomes?

The answer became the foundation of self-determination AI.

A metaphor I often use with students and executives is the difference between a fruit picker and a digital garden owner. It captures a dynamic that has quietly shaped the digital economy for years.

Millions of people spend their days producing value inside platforms they do not own. They create content for social networks, data for corporations, and intellectual labor for systems that capture most of the economic benefit. They are productive, but they are not owners.

AI magnifies this divide. Those who simply use AI tools may become more efficient. Those who learn to orchestrate AI systems may become owners of entirely new forms of value.

A digital garden is not a website or a brand. It is a portfolio of assets, including knowledge, systems, data, relationships, and intellectual property, cultivated intentionally over time. It grows through curiosity, experimentation, and strategic design.

A fruit picker works hard but owns nothing. A digital garden owner builds assets that compound. The purpose of self-determination AI is to help people make that transition.

The Four Pillars of Ownership

Self-determination AI rests on four forms of ownership. Together, they create a foundation for human agency and strategic advantage in an intelligent world.

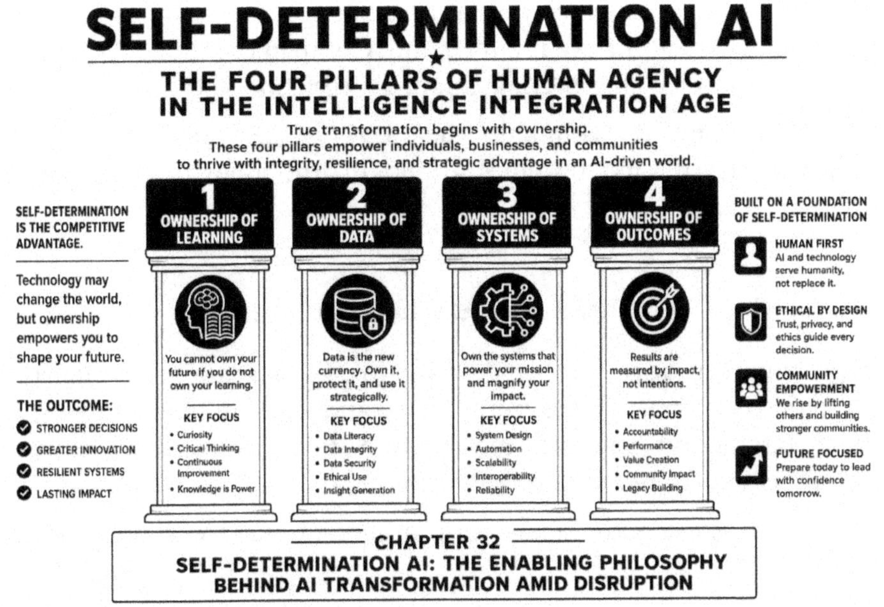

Ownership of Learning: The New Competitive Advantage

Learning has always mattered, yet in the AI era, it has become the primary driver of opportunity. When knowledge evolves rapidly and industries shift overnight, the ability to learn quickly becomes more valuable than any specific skill.

AI accelerates learning, but only for those who take responsibility for the process. Some people use AI as a shortcut. They ask for answers and accept whatever it provides. Others use AI as a catalyst. They ask better questions, explore deeper ideas, and use technology to expand their thinking rather than replace it.

Ownership of learning means treating curiosity as a strategic asset. It means using AI to explore, not to avoid effort. It means building a personal knowledge system and developing the ability to synthesize, evaluate, and apply information. It means recognizing that learning is no longer episodic, but continuous.

Organizations face the same challenge. Companies that treat learning as a one-time event fall behind. Companies that build cultures of continuous learning become adaptive, resilient, and innovative.

AI does not eliminate the need for learning. It increases it. When information becomes abundant, the ability to interpret it becomes scarce.

Ownership of Data: The New Digital Soil

In previous centuries, land created wealth. Today, data plays that role. It shapes decisions, drives innovation, and fuels intelligent systems. Every interaction generates information. Every transaction leaves a trail. Every customer creates a pattern.

Yet most individuals and organizations treat data as an afterthought.

Ownership of data means understanding what information you generate, how it is used, and how it can support your goals. It means curating datasets that reflect your work, your insights, and your value. It means protecting your privacy and maintaining autonomy in a world where algorithms increasingly influence opportunity.

For individuals, this might involve building a personal knowledge archive, tracking your learning, or analyzing your own productivity patterns. For organizations, it means treating data as a strategic asset rather than a technical byproduct.

Data is not just a resource. It is a form of identity. Those who take ownership of it gain leverage. Those who ignore it become invisible.

Ownership of Systems: The Architecture of Leverage

Knowledge creates potential. Systems convert that potential into leverage.

AI enables individuals and organizations to build systems that automate tasks, streamline workflows, and operate continuously.

These systems can generate content, analyze information, manage communication, and support decision-making. Using AI is helpful. Designing AI-enabled systems is transformative.

Ownership of systems means creating environments where technology works on your behalf, even when you are not actively engaged. It means building processes that scale your expertise, extend your reach, and multiply your impact.

For organizations, this is where competitive advantage emerges. Companies that build AI-enabled systems reduce friction, accelerate innovation, and create compounding value. They move from manual effort to strategic orchestration. Systems turn knowledge into capability, capability into advantage, and advantage into scale.

Ownership of Outcomes: The Discipline of Value Creation

Activity is not the same as achievement. Automation is not the same as progress. AI can generate output at extraordinary speed, yet only people can determine whether that output creates meaningful value.

Ownership of outcomes requires clarity. It demands intentionality. It asks individuals and organizations to define success, measure progress, and align action with purpose.

In practice, this means setting clear goals, evaluating results honestly, building assets that compound over time, and ensuring that technology serves your objectives, not the other way around.

Organizations that confuse activity with outcomes become busy but ineffective. They automate processes without improving performance. They generate data without generating insight. They produce content without producing value.

Ownership of outcomes is the antidote to digital noise.

Self-Determination AI as the Enabling Strategy for Thriving in the AI Transformation

The integration of a self-determination AI ownership strategy shifts attention away from the technology's capabilities and toward the amplified capabilities of the people who orchestrate it. This shift represents a fundamental reorientation of power. Thriving in the next major disruption requires moving from passive participation in digital platforms, the equivalent of a fruit picker working someone else's land, to becoming an active producer and owner of a personal digital estate, a digital garden cultivated with intention.

This strategy strengthens human potential by grounding it in four essential forms of ownership. Each pillar aligns with a core psychological need identified in self-determination theory, and each one expands a person's capacity to act with clarity and purpose in an increasingly automated world.

Amplified Human Capabilities Through Ownership

Ownership of Learning (Competence)

The first pillar is commitment to lifelong literacy and the ability to learn new things quickly. In the AI era, this is not merely an advantage. It is the ultimate differentiator. Research consistently shows that when individuals feel competent with emerging tools, their optimism toward AI increases, their anxiety decreases, and their engagement deepens. Mastery fuels momentum. People who take ownership of their learning build the confidence required to navigate complexity, adapt to new environments, and create value in ways that machines cannot replicate.

Ownership of Data (Autonomy)

Data has become the digital soil of the modern economy. Treating personal and professional data as a strategic asset gives individuals the autonomy to act in alignment with their own values rather than being shaped, often invisibly, by external algorithms. When people understand the information they generate, how it is used, and how it can be leveraged, they reclaim agency in a world where data often flows in one direction, away from the individual and toward the platform. Ownership of data restores balance.

Ownership of Systems

Most people interact with technology at the surface level, through apps, interfaces, and tools designed by others. Ownership of systems invites individuals to go deeper. Instead of relying on isolated applications, they learn to orchestrate agentic AI workflows that operate on their behalf. This is where autonomy becomes tangible. When people design systems that reflect their goals, their values, and their priorities, technology stops being a source of overwhelm and becomes a force multiplier.

Ownership of Outcomes

The final pillar reframes success. Instead of measuring progress by activity or effort, individuals begin building independent, income-producing systems and compounding assets that grow in value over time. Ownership of outcomes encourages people to think beyond tasks and toward long-term value creation. It shifts the focus from productivity to prosperity.

Impact on Thriving Within Digital Disruption

The impact of this ownership strategy on an individual's ability to thrive during massive disruption is profound.

From Dependency to Independence

Most people experience technology the way sailors experience an iceberg, aware of the visible portion but unaware of the vast structure beneath the surface. The underlying design, incentives, and control mechanisms remain hidden. A self-determination AI strategy dives beneath the waterline. It enables individuals to understand, influence, and ultimately design the digital systems that shape their lives. Independence begins with visibility.

Psychological Resilience

Decades of research in self-determination theory show that when the basic psychological needs for autonomy, competence, and relatedness are supported, individuals experience lower anxiety, greater motivation, and higher well-being. The same holds true in the AI era. People who feel capable of learning new tools, understand their data, and can design their own systems report significantly lower levels of AI negativity. They adapt intentionally rather than reactively. They remain grounded even as the world accelerates around them.

The Wisdom–Automation Boundary

AI can automate thinking, yet it cannot automate understanding. It can generate content, yet it cannot generate purpose. It can identify patterns, yet it cannot determine what matters. A self-determination strategy reinforces this boundary. It reminds individuals that while they can outsource tasks to AI, they cannot outsource judgment. Thriving depends on maintaining this human edge, the ability to interpret, evaluate, and decide.

The Wisdom Boundary: Where Humans Remain Essential

As AI becomes more capable, a boundary emerges: the line between intelligence and wisdom. AI can process information, yet it cannot determine meaning. It can identify patterns, yet it cannot define purpose. It can generate content, yet it cannot generate values. This boundary is where human beings remain indispensable.

The future will not be shaped by machines alone, nor by humans who refuse to adapt. It will be shaped by those who understand how to combine the speed and scale of AI with the judgment, ethics, and purpose of human beings.

Self-determination AI encourages people to use AI for acceleration rather than as a substitute. It reinforces the importance of human judgment. And it reminds us that while we can outsource tasks, we cannot outsource understanding.

Thriving in Civilization's Next Big Disruption

The title of this book includes a word that deserves attention: thriving. Survival is reactive. Thriving is intentional.

AI will reshape industries, redefine professions, and transform institutions. Some roles will disappear. New opportunities will emerge. Yet the individuals and organizations that thrive will not be those with the most advanced tools. They will be those with the greatest capacity for ownership.

These pillars transform individuals from passive consumers into active producers. They shift people from dependency toward independence. They move individuals from being fruit pickers to becoming digital garden owners.

Artificial intelligence amplifies capability. Ownership amplifies humanity. Those who learn to combine both will not merely survive civilization's next big disruption. They will shape it.

Artificial intelligence will undoubtedly change how we work, learn, communicate, and create value. Yet the most important question remains unchanged: Will we become more dependent on technology, or will we use technology to become more capable?

Self-determination AI offers a pathway toward the latter.

The future belongs to those who learn continuously, manage information wisely, build systems intentionally, and create outcomes that matter. Technology may amplify capability, but ownership amplifies humanity.

That is the central message of this chapter, and it is why I believe self-determination AI represents an important contribution to understanding how individuals and organizations can thrive amid civilization's next big disruption.

About the Author

Professor Gary Vause, II, is an international scholar-practitioner, educator, entrepreneur, US Air Force veteran, and technology strategist. He is CEO of Vause Computer Systems, LLC, a consulting firm specializing in digital transformation, cybersecurity, data analytics, artificial intelligence, and intelligence integration.

A former KPMG Advisory professional, Professor Vause has spent more than three decades helping organizations navigate technological change and business transformation. His professional experience includes consulting engagements with government agencies, Fortune 500 companies, educational institutions, healthcare organizations, and law enforcement agencies.

He is the creator of self-determination AI, a framework built upon ownership of learning, ownership of data, ownership of systems, and ownership of outcomes. His research explores how individuals, businesses, and communities can thrive in the Intelligence Integration Age by combining human judgment, continuous learning, and intelligent technologies.

Professor Vause holds degrees in psychology, computer science, behavioral science, education, public administration, and technology science. He is a US Air Force veteran, international lecturer, and co-author of *The AI Transformation: Thriving Within Civilization's Next Big Disruption.*

Email: gvause2@gmail.com

Website: www.GaryVause2.com

LinkedIn: https://www.linkedin.com/in/gary-vause2/

CHAPTER 33

THE AI-ENABLED ENTERPRISE: THRIVING IN THE NEVER NORMAL

By Rick Wilson
Founder, Digitalruption Inc.
Brooklyn, New York

We are moving from a world of episodic change to one of continuous change. The new normal is that there is no normal.
—Peter Hinssen (author of *The Never Normal*)

The idea of "normal" has always been a convenient fiction, an anchor point that allows individuals and organizations to orient themselves in a world of change. But in the age of artificial intelligence, that anchor is dissolving. What we are experiencing is not simply another wave of technological disruption; it is a fundamental shift in how intelligence itself is created, distributed, and applied.

In *Never Normal*, the premise is clear: Change is no longer episodic. It is continuous. AI accelerates this reality to an entirely new level. It introduces a world where learning is exponential, adaptation is constant, and the boundary between human and machine intelligence becomes increasingly fluid. In this environment, thriving is not about mastering a new toolset; it is about rethinking how we live, work, lead, and evolve.

This chapter explores what it means to operate in this "never normal" state in the context of AI and how individuals and enterprises can cultivate the mindset, structures, and capabilities required to succeed.

From Digital Transformation to Organizational Intelligence

For the past two decades, organizations have invested heavily in digital transformation. Enterprise systems, cloud computing, customer platforms, and data infrastructure have reshaped the technological backbone of business. Yet despite these investments, many organizations remain fundamentally static in how they operate.

The problem is not technology. It is about how technology is integrated into the organization's intelligence.

Traditional transformation focuses on implementation: deploying systems, automating processes, and digitizing workflows. But AI introduces a different paradigm, one that shifts the focus from systems to intelligence. The question is no longer "What technology have we implemented?" but "How intelligently does our organization sense, interpret, and act?" This shift can be understood through a simple but powerful model:

Signal → Interpretation → Decision → Action

In a conventional enterprise, these steps are slow, fragmented, and often disconnected. Data is collected but not fully understood. Decisions are delayed by hierarchy. Actions are constrained by rigid processes.

In an AI-enabled enterprise, this loop becomes continuous and dynamic. Signals are captured in real time from customers, employees, markets, and operations. AI systems assist in interpreting those signals, surfacing patterns and insights that would otherwise remain hidden. Decisions are augmented with predictive modeling and scenario analysis. Actions are executed through intelligent workflows that adapt as conditions change. The organization, in effect, becomes a living system.

This is the essence of organizational intelligence: the ability to continuously evolve in response to the signals it receives. Competitive advantage shifts away from owning technology and toward designing intelligence architectures, systems that enable perpetual sensing, learning, and adaptation.

Redesigning Work: From Tasks to Intelligent Flows

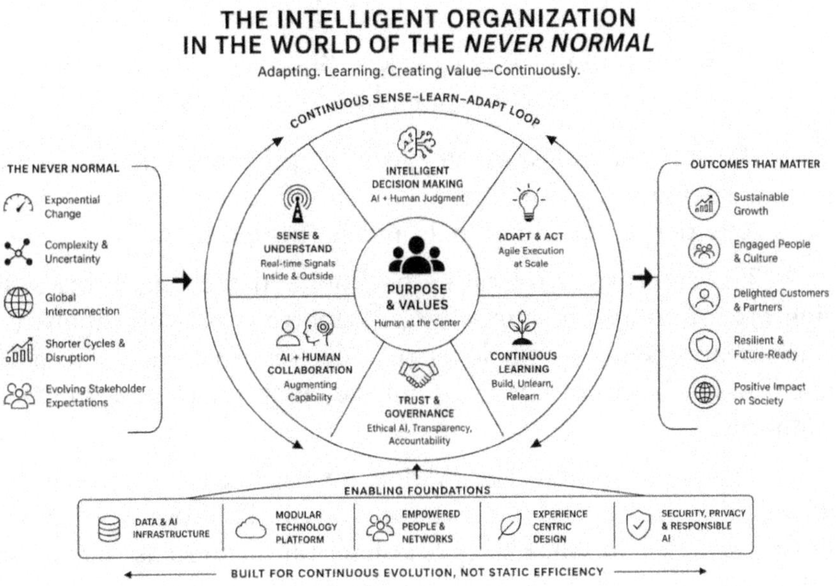

One of the most profound impacts of AI is on the nature of work itself. For over a century, work has been structured around roles, tasks, and hierarchies. Job descriptions define responsibilities. Processes dictate how work is performed. Organizational charts determine authority.

AI disrupts this model at its core. Rather than organizing work around static roles, AI enables the orchestration of dynamic flows. Tasks are no longer owned exclusively by individuals; they are distributed across human and machine capabilities. AI agents can handle routine decisions, route work based on real-time conditions, and manage exceptions with increasing sophistication.

This does not eliminate the need for humans. It redefines their value. In an AI-enabled environment, human contribution shifts toward areas where machines are least effective:

- Judgment: Navigating ambiguity, ethics, and context
- Exception Management: Handling the unexpected and the non-linear
- Creativity: Generating novel ideas and solutions
- Relationship Intelligence: Building trust, influencing outcomes, and understanding human nuance

Work becomes less about executing predefined tasks and more about participating in intelligent systems. Individuals move fluidly across activities, guided by signals and supported by AI.

The enterprise's operating backbone also changes. Instead of being organized around functions or departments, work aligns with value streams: end-to-end flows that deliver outcomes. AI enables these flows to be continuously optimized, adapting to new inputs and conditions.

The implication is clear: AI does not replace work. It repositions human value within it. Those who thrive will be those who embrace this shift and develop the ability to collaborate with AI as an extension of their own capabilities.

Rewiring Leadership: From Control to Signal Interpretation

If work is being redefined, leadership must evolve as well. Traditional leadership models are built on control: setting direction, making decisions, and ensuring execution. This approach assumes a relatively stable environment where information flows upward, decisions flow downward, and change occurs at a manageable pace.

In a "never normal" world, this model breaks down. The volume, velocity, and variability of information exceed what any individual or hierarchy can manage. Decisions must be made in real time, often with incomplete information. The environment is too complex for centralized control. Leadership, therefore, shifts from control to interpretation.

The modern leader is not simply a decision-maker; they are a navigator of signals. Their role is to:

- Detect patterns in a sea of data
- Interpret ambiguity and uncertainty
- Integrate insights from multiple sources (human and AI)
- Enable rapid, informed action

AI becomes a critical partner in this process. It enhances leaders' ability to model scenarios, assess risks, and explore potential outcomes. But it does not replace leadership. It amplifies it.

Perhaps most importantly, leadership becomes less about authority and more about alignment. In a dynamic system, alignment cannot be imposed; it must be continuously achieved. This involves managing a cycle of:

Divergence → Visibility → Convergence → Cohesion

- **Divergence:** Different perspectives and ideas emerge.
- **Visibility:** These perspectives are surfaced and understood.

- **Convergence:** Insights are synthesized into direction.
- **Cohesion:** The organization aligns and acts.

AI can support each stage, but leaders must orchestrate the process. They must create environments where signals are visible, dialogue is encouraged, and decisions are informed by both data and human insight.

The leader of the future is not the one with all the answers but the one who can integrate intelligence across the system.

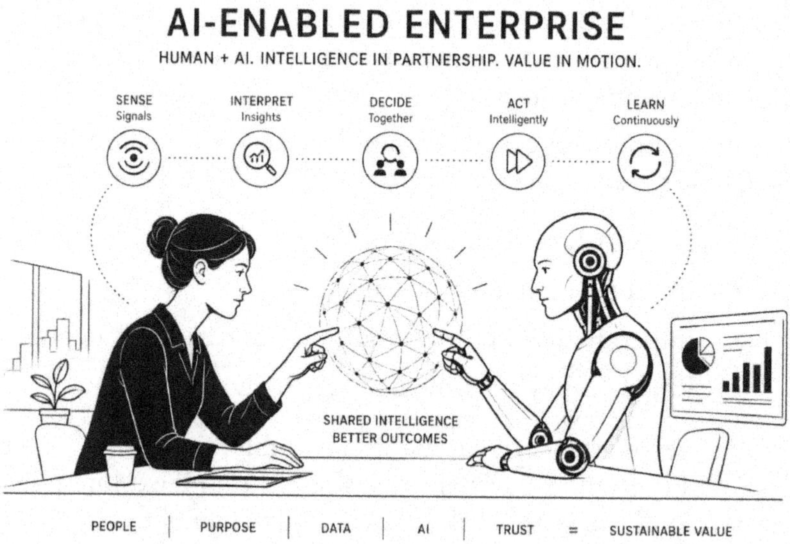

Reinventing Learning: The Living Capability System

If AI changes how we work and lead, it also transforms how we learn.

Traditional learning models are built on periodic training: courses, workshops, and certifications delivered at specific points in time. These approaches assume that skills can be acquired, retained, and applied over extended periods.

In a world of exponential change, this assumption no longer holds. Skills become obsolete quickly. New capabilities emerge continuously. The gap between learning and application becomes a critical barrier to performance.

AI enables a new model: the living capability system. At its core is a continuous loop: **Sense Gaps → Activate Learning → Measure → Refine**

- **Sense Gaps:** Identify where capabilities are lacking, based on real-time performance and signals
- **Activate Learning:** Deliver targeted, contextual learning experiences
- **Measure:** Assess impact in terms of performance and outcomes
- **Refine:** Adjust learning approaches based on feedback and results

Learning becomes embedded in the flow of work. Instead of stepping away to learn, individuals learn as they perform. AI systems provide guidance, recommendations, and feedback in real time, tailoring experiences to individual needs and contexts.

This shift has profound implications for culture. Learning is no longer an event. It is a constant state. Organizations move from a mindset of "training people" to one of "engineering capability." The focus is not on delivering content but on enabling performance.

Individuals, in turn, must adopt a new relationship with learning. It becomes less about acquiring knowledge and more about continuously evolving. Curiosity, adaptability, and self-directed growth become essential traits.

In the "never normal," the ability to learn is not a supporting capability. It is the core capability.

Designing the AI-Enabled Operating Model

To fully realize the potential of AI, organizations must rethink their operating models. This is not about layering AI onto existing structures; it is about redesigning the enterprise as an intelligent system. One way to conceptualize this is through five interconnected layers:

1. **AI Infrastructure Platform**

   The foundational technologies—cloud, data architecture, machine learning models—that enable AI capabilities

2. **Business Growth Signals**

   The mechanisms for capturing and interpreting signals from the market, customers, and operations

3. **Capability Intelligence**

   The systems that understand and develop human and organizational capabilities

4. **Hybrid Workforce Network**

   The integration of human and AI agents into a cohesive system of work

5. **Organizational Experience Design**

   The orchestration of experiences across employees, customers, and stakeholders

Together, these layers form a dynamic ecosystem.

What distinguishes this model is its emphasis on experience and adaptability. The organization is not a static structure, but a continuously evolving system shaped by interactions between humans and AI. Experience becomes the unifying element. Employee experience, customer experience, and user experience converge with organizational design to create a holistic environment. Decisions about structure, process, and technology are guided by how they enhance the overall experience.

This approach also reflects a deeper shift: from efficiency to intelligence. Traditional operating models prioritize efficiency: doing things faster and cheaper. AI-enabled models prioritize intelligence: doing things more effectively based on better understanding. Efficiency still matters, but it is no longer sufficient. In a world of constant change, the ability to adapt intelligently becomes the primary source of value.

Living in the Never Normal

While much of this chapter has focused on organizations, the implications extend to individuals and society as a whole. Living in the "Never Normal" requires a fundamental shift in mindset. First, it demands comfort with uncertainty. AI accelerates change to the point where predictability becomes limited. Instead of seeking stability, individuals must develop the ability to navigate ambiguity.

Second, it requires a redefinition of identity. In a world where roles are fluid and capabilities are constantly evolving, identity cannot be tied to a specific job or skillset. It must be rooted in adaptability, purpose, and the ability to learn.

Third, it calls for a new relationship with technology. AI is not just a tool. It is a collaborator. Those who treat it as such will unlock far greater value than those who view it merely as automation.

Finally, it emphasizes the importance of humanity. As AI becomes more capable, the uniquely human aspects of work and life become more valuable. Empathy, ethics, creativity, and relationships are not diminished by AI. They are amplified in importance.

The challenge, therefore, is not to compete with AI but to integrate it in ways that enhance human potential.

The Cultural Shift: From Stability to Evolution

At the heart of all these changes is culture. Culture determines how organizations respond to change, whether they resist it, adapt to it, or leverage it. In the age of AI, culture must evolve from one that values stability to one that embraces continuous evolution. This involves several key shifts:

- **From Certainty to Curiosity:** Encouraging exploration and questioning
- **From Control to Empowerment:** Enabling individuals to act on insights
- **From Hierarchy to Networks:** Facilitating collaboration across boundaries
- **From Static Roles to Dynamic Capabilities:** Supporting fluid movement and growth

Building such a culture is not easy. It requires intentional design, leadership commitment, and alignment across the organization. But it is essential.

Without the right culture, even the most advanced AI technologies will fail to deliver their potential. With the right culture, organizations can harness AI to create entirely new forms of value.

The Intelligent Evolution

The AI transformation is not a destination. It is an ongoing journey. Organizations and individuals are not simply adopting new technologies; they are participating in the creation of a new kind of intelligence, one that emerges from the interaction of humans and machines.

In this context, the "never normal" is not something to be feared. It is the natural state of a world where intelligence is continuously expanding.

The question is not whether we can return to normal. We cannot. The question is whether we can evolve. Those who do will not be defined by what they build, but by how intelligently they adapt. They will design systems that learn, cultures that evolve, and experiences that inspire. They will not merely survive the disruption. They will thrive within it. And in doing so, they will redefine what it means to be human in an age of artificial intelligence.

About the Author

For more than two decades, Rick Wilson has worked at the intersection of organizational design, workforce transformation, operating model strategy, enterprise change, digital transformation, and artificial intelligence. His work focuses on moving beyond persistent, episodic digital transformation toward building agile and continuously flowing intelligent organizations — where human capability, AI, organizational design, stakeholder alignment, and decision-making operate in a unified ecosystem.

LinkedIn: https://www.linkedin.com/in/rickwilson-ocm/

DID YOU ENJOY THIS BOOK?

If you enjoyed reading this book, you can help by suggesting it to someone else you think might like it, and **please leave a positive review** wherever you purchased it. This does a lot in helping others find the book. We thank you in advance for taking a few moments to do this.

THANK YOU

You might also like other Thin Leaf Press titles:

The AI Advantage: Thriving Within Civilization's Next Big Disruption

The AI Revolution: Thriving Within Civilization's Next Big Disruption

The AI Mindset: Thriving Within Civilization's Next Big Disruption

AI: Work Smarter and Live Better Within Civilization's Next Big Disruption

Peak Performance: Mindset Tools for Managers

Peak Performance: Mindset Tools for Sales

Peak Performance: Mindset Tools for Leaders

Peak Performance: Mindset Tools for Business

Peak Performance: Mindset Tools for Entrepreneurs

Peak Performance: Mindset Tools for Athletes

The Successful Mind: Tools to Living a Purposeful, Productive, and Happy Life

The Successful Body: Using Fitness, Nutrition, and Mindset to Live Better

The Successful Spirit: Top Performers Share Secrets to a Winning Mindset

Winning Mindset: Elite Strategies for Peak Performance

Winner's Mindset: Peak Performance Strategies for Success

The Life Coach's Tool Kit, Vol. 1

The Life Coach's Tool Kit, Vol. 2

The Life Coach's Tool Kit, Vol. 3

Ordinary to Extraordinary

The Magical Lightness of Being

Explore.

www.ingramcontent.com/pod-product-compliance
Lightning Source LLC
LaVergne TN
LVHW021740071126
851857LV00039B/573
* 9 7 8 1 9 6 8 3 1 8 7 3 4 *